An Ecocritical Study of Puerto Rican Culture

María Acosta Cruz

An Ecocritical Study of Puerto Rican Culture

Disaster Nation

palgrave
macmillan

María Acosta Cruz
Language, Literature and Culture
Clark University
Worcester, MA, USA

ISBN 978-3-032-04043-5 ISBN 978-3-032-04044-2 (eBook)
https://doi.org/10.1007/978-3-032-04044-2

This Palgrave Macmillan imprint is published by the registered company Springer Nature Switzerland AG
The registered company address is: Gewerbestrasse 11, 6330 Cham, Switzerland

Acknowledgments The Alice Coonley Higgins Institute for Arts and Humanities at Clark University has been invaluable for my intellectual development. Its director, Professor Matt Malsky, and Associate Director, Jennifer McGugan, fully supported this project. I have been so lucky to have been a part of the Institute since its inception.

Dr. Miguel Angel Náter, director of the Universidad de Puerto Rico-Río Piedras's indispensable Seminario Federico de Onís generously shared his research expertise. The Instituto de Cultura Puertorriqueña kindly facilitated use of their databases and library, as this entire book shows.

Dr. Chris Landsea, Branch Chief of the Tropical Analysis and Forecast Branch from NOAA/NWS/NCEP/National Hurricane Center directed me to climate historian Michael Chenoweth, who unstintingly shared his world-class expertise about hurricanes.

My heartfelt gratitude to Clark University colleagues, particularly Professor Juan Pablo Rivera, whose enthusiasm, insightful feedback and thoughtful suggestions enrich my thinking.

Thanks to my indexer, Mixie Cesarano, whose meticulous attention to detail ensured that the index is accessible and well-organized. And to my youngest daughter, June Cesarano, unpaid research assistant, couldn't have done it without you.

Camille Davies, Senior Editor at Palgrave Macmillan, was a steady hand guiding the book to fruition. Agila Muruganandam, Production Editor for Springer, answered any and all questions.

Finally, I thank my thoughtful husband, the Honorable William H. Abrashkin, whose love, support, and editing skills made Chapter 1 a reality.

Competing Interests The author has no competing interests to declare that are relevant to the content of this manuscript.

Contents

List of Figures

CHAPTER 1

Introduction

De la tormenta bravía

del escollo misterioso,

líbrame, cielo amoroso

("From the fierce storm

from the mysterious shoal,

deliver me, loving heaven")

"La plegaria de una virgen" ("A Virgin's Prayer")

Alejandro Tapia y Rivera (*Obras* "Works" 1968; 567)

Literary and artistic depictions of the environment are never innocent. How a nation becomes a Nation depends to a large extent on how it imagines, narrates, symbolizes and, in sum, interprets its environment. In *Disaster Nation* we use ecocriticism (combining science and the humanities) to explore how disasters have been represented (or not, and if so why) in Puerto Rican literature, history and the arts.

M. Acosta Cruz, *An Ecocritical Study of Puerto Rican Culture*,
https://doi.org/10.1007/978-3-032-04044-2_1

We research the construction of a national imagination in Puerto Rico, a nation with unique convergences of calamities: cyclonic location and geological instability, as well as continuous colonialism. Using ecocriticism (combining scientific and historical data) we delve into the themes and threads in cultural production centered on representations or lack thereof (ecoerasure) of hurricanes and earthquakes throughout the span of five plus centuries, from Taino times to 1508 (the first recorded hurricane) and up to the aftermath of the 2020 earthquakes.

Puerto Rico's culture is a complex web of references to disasters. Sometimes, Puerto Rican history, literature and arts highlight the drama of hurricanes, earthquakes, and the governmental inefficiencies that heighten them. For example, Enrique Laguerre's 1940 novel *Solar Montoya* has a hurricane-driven plot, and José Campeche's painting, *San Emigdio de Áscoli*, references the earthquakes of 1787. But time and again the classics we read in universities and gaze at in museums display a placid landscape, an Edenic garden of eternal spring.

Puerto Rican literature and history have created works that seesaw from stark descriptions of hurricanes and earthquakes to a fictionally serene Nature, and then back to describing the trauma and resilience of people confronting an often-violent Nature. Meanwhile, art has taken more linear paths, from very few portrayals of disasters (i.e., José Campeche), to depictions of placid, bucolic landscapes, and now to a proliferation of images of the paths of disaster, especially post-hurricane Maria.

Using ecocritical tools, this book situates each work of culture within its environmental conditions, documenting crucial disasters in years prior to the creation of that work and also during the creator's lifetime. In this manner we assess the proximity of the stories, images, and symbols of the environment to Puerto Rico's climate and geologic reality (ecorealism) as we explore notions of nationhood, disaster, trauma, and resilience. Our ecocritical tools are: interdisciplinarity—combining meteorological and historical data—ecofeminist analyses, close readings, and postcolonial scrutiny centering on environmental justice. We know representations of the landscape in any culture undergird a country's history; and yes, Puerto Rican writers have also focused on matters of race and gender, labor, sovereignty, social justice, and colonial extractivism, but much of its culture either grapples with or denies its Caribbean environment.

This book follows the links between culture and politics in its author's book *Dream Nation*, which studies the mechanisms of a culture that has always thrived on one over-arching cultural directive: stories and images linked to the dream of independence. Here we offer an environmental analysis equally challenging because we regard the Nation as having solidified around ecoerasing, ahistorical views of the natural environment.

The historical and scientific data in this book has never been collected in one place before. Our Appendix X lists the main characteristics of all major disasters to have struck the Puerto Rican archipelago alongside the dates of major works of culture. In this way we stress the importance of the scientific and historical background of hurricanes and other disasters and equally, how environmental degradations are due to colonialism (Concepción 1988, 128).

We trace how stories, symbols, and images can transform disasters into memorable and compelling events about the human and non-human world. Furthermore, we explore how artistic and literary responses to disasters can affect preparedness and resilience. For instance, twentieth-century art and literature increasingly stressed disaster preparedness–especially the art from DIVEDCO, a government agency that focused on preparedness education. Twenty-first century art, history, and literature also passionately explore the ways community-led climate activism and planning for resilience impact life and culture in Puerto Rico, as disaster capitalism continues unabated and the planet's oceans warm up.[1]

Key Concepts and Recursively-Deployed Terms Used in This Book

First is our term "ecoerasure," the cultural forgetting or erasing of an inherently violent environment. The term joins Simon E. Estok's term "ecophobia" (contempt or disregard for the natural environment) (2010, 141) and its opposite, Ruyu Hung's notion of "ecophilia" (human appreciation of other living beings and their environments) (2017, 45). Ecoerasure, ecophobia, and ecophilia undergird the Puerto Rican Nation throughout five hundred plus years of cultural production, as this book shows.

Another key term is "hurricane," or cyclone in the Atlantic Ocean, the meteorological term for a violent counterclockwise rotating system of clouds and thunderstorms over tropical or subtropical waters bringing sustained winds of 74 mph or higher. Hurricanes in the Atlantic rotate counterclockwise and are (so far) classified on the Saffir-Simpson Scale in categories 1 through 5 [2] based on maximum sustained wind speed. Typhoons are cyclones in the Pacific Northwest.

Next is "ecocritism," a field of culture and ecology that uses critical tools from both the humanities and the sciences. We deploy ecocriticism to study the relationship between literature and the physical environment. William Rueckert coined the term in 1978 and it became an important tool of cultural analysis during the 1990s. The publication in 2023 of the third edition Greg Garrard's *Ecocriticism* reaffirms the endurance of this critical movement.

Equally important for us is material feminism, which stresses the effects of climate change for both humans and non-humans.

The word "Nature" (or Mother Nature, always capitalized), is part of an ecophilic tradition associated with the Romantic era. The totality of the natural world comes under the synonyms "nature" and the "environment" (here employed interchangeably). Boricua scholars are cited throughout the book, linking our literary and artistic analyses to environmental justice, therefore framing our views on culture within the need for climate and energy justice in Puerto Rico.

"Disaster" refers to traumatic catastrophes (natural or human-created) that impact large numbers of people. We support current scholarship's critiques of the word "natural" in "natural disasters" [3] as the present book uses the designation "Capitalocene" for how colonialism and capitalism have negatively impacted climate. Extractivist capitalism remains one of the most widespread of disasters.

Finally, "paleotempestology" documents the long-term fluctuations in hurricane activity before the existence of written records (cf. Fan and Liu 2008). This definition and its uses are expanded in Chapter 2.

As Taino mythology and ceramics show, Borinquen's culture has always dealt with disasters since well before Juan Ponce de León first documented a hurricane in 1508. The early modern Spanish chroniclers would often focus on hurricanes and earthquakes, with particular emphasis on religious interpretations of Nature as well as tales of manly colonial heroes battling Nature. These accounts were usually beholden to the colony's geopolitical situation. Nevertheless, later, during its nation-building period (starting in the late eighteenth century), Puerto Rico's culture portrays Nature as benevolent, with fictions and images of a landscape untroubled by devastating winds, catastrophic flooding, or wrecking earthquakes. These placid depictions follow an understandable aversion to dwelling on climate trauma and are inspired by patriotic love of

Land. Given the real political disempowerment that Puerto Rico suffers, ecoerasing love of a pristine old-timey *Isla* enhanced the archipelago's emerging national identity. Our Appendix X documents how ecoerasure also flourished during gaps in hurricane activity[4]; pauses that enabled what Mike Davis labels "disaster amnesia" (1999, 47), despite the well-documented scars of disaster in Puerto Rico.

As Wilber W. Caldwell says, all nations lean toward exceptionalism, a "universal characteristic of nationalism," feelings that are heightened by "anger, sentimentality, fear" (2006, 2). Puerto Rican exceptionalism (the belief that its virtues are unique) imagines a pristine land with the most beautiful, unspoiled beaches, the loveliest women, the most delicious food, and only good things, never mentioning disasters pummeling the land. Though sentimental, however, Puerto Rican exceptionalism is equally a defensive response to ongoing foreign dominion. How does exceptionalism deal with hurricanes and their aftermath? With silence.

Exceptionalism in what amounts to a Hurricane Alley overwhelmingly ignores the situation of a Caribbean region where climate change brings increasing real-life challenges to island nations. The dangers are many because, "islands are hotspots of cultural, biological and geophysical diversity" (Kueffer and Kinney 2017, 311). Specific impacts of climate change are outlined in the United Nations' "Intergovernmental Panel on Climate Change (IPCC) Sixth Assessment Report" which asserts the interdependence of climate, ecosystems and biodiversity, and human societies. Puerto Rico experiences many of the effects the report summarizes: "biodiversity loss, overall unsustainable consumption of natural resources, land and ecosystem degradation, rapid urbanisation, human demographic shifts, social and economic inequalities" (Pörtner 2020, 5). Specifically, the report states human-induced climate change increases the frequency of tropical hurricanes. These climate disruptions degrade infrastructure and imperil public health, causing, for example, the morbidity following Hurricane Maria. Many of the conditions in the U.N. lists are present in the Puerto Rican archipelago, where people's vulnerability differs according to socio-economic inequities caused by colonialism and marginalization, as well as unsustainable environmental abuse (Pörtner 2020, 12).

What's in a Name? The cultural impact of certain names like San Roque, San Ciriaco, San Felipe and, now, Maria, make hurricanes crucial characters in the story of Puerto Rico. That is why our Appendix X catalogues weather events ranked as hurricanes, i.e., those that have names, although tropical storms and nameless *vaguadas* can cause extensive damage, with rainfall that destroys electric power grids, water supply systems, and infrastructures.[5]

We use ecocriticism in relation to socio-political issues and climate justice to deploy a core ecocritical premise that "environmental crises have social, cultural, affective, imaginative, and material dimensions" (Lousley 2020). This means that here we examine how culture interacts with the real-life consequences of disaster, examining historical and scientific documentation of the effects of each catastrophe. This book deploys ecocriticism as: "the study of the relationship between literature and the physical environment. Just as feminist criticism examines language and literature from a gender-conscious perspective, and Marxist criticism brings awareness of modes of production and economic class to its reading of texts, ecocriticism takes an earth-centered approach to literary studies" (Glotfelty and Fromm 1997, xix).

Puerto Rican patriarchal representations of women link them to the landscape where often they mirror a fictionally placid environment (granting them little or no agency), or they replicate cyclones (destructive forces beyond control or understanding). This book stresses ecofeminism [6] to signal the interconnectivity between heteronormative visions of "woman" and "Nature"—from the Romantic trope of an "isla doncella" ("maiden isle") to Luis Palés Matos's wild hurricane women to Julia de Burgos's activist ecopoetics of the aquatic environment. We see this burden of representation for women even in pop culture, like the video of the song "Despacito," in which Miss Universe, Zuleyka Rivera, is the equal of a palm tree. As Stacy Alaimo says, "'woman' has long been defined in Western thought as being mired in 'nature' and thus outside the domain of human transcendence, rationality, subjectivity, and agency" (2008, 239).

We agree that stories, images and symbols can be about, "not only sentient animals or other biological organisms, but also impersonal agents, ranging from electricity to *hurricanes*" (Iovino and Oppermann 2014, 3; our emphasis). The main premise of material feminism is that the human and the non-human have agency and that the environment is composed of entities that act, influence, or control a setting, and therefore should be featured as such in culture. Agency is not exclusive to human intentionality, but is "a pervasive and inbuilt property of matter, part and parcel of its generative dynamism" (Iovino and Oppermann 2014, 3). Hannes Bergthaller goes further, stating that matter has agency in that it is "endowed with many of the same qualities that were formally seen as exclusive to human beings: complex self–organization, reflexivity, consciousness, and the capacity to act spontaneously, that is, in a manner not reducible to external determination" (2014, 37). Elizabeth A. Povinelli clarifies intentionality, that is to say, the question of the ontology of the living versus the non-living, when she defines agency as "the force of one thing to cause something else to reformulate its course of action" (2015, 437).

We understand hurricanes to have agency in Puerto Rico's culture because they have driven cultural production in specific directions. Fernando Ortíz the great Cuban anthropologist, says hurricanes are characters with the equivalent of free will because they are erratic characters (1947, 50). And, as Lilliana Ramos Collado says, hurricanes are monsters (2012, 7). This imaginative background sustains our treatment of hurricanes throughout the book, because we consider these cataclysms as crucial characters in the story of the Nation.

Hurricanes churn up coastlines as well as culture and therefore have agency, as writers like Luis Palés Matos, artists like Rafael Tufiño and historians like Luis A. Salivia prove. Tales and images of migration, for instance, are often reactions to the uncontrollable assault of hurricanes and earthquakes. Storm stories vary because different trajectories generate vastly disparate stories and images.[7] Hurricanes cause climate anxiety because of the vast ground they cover, typically about 300 miles wide. Technology accurately follows hurricanes, but their devastation is never foreseeable. For instance, more than 300 people died during Category 5 San Felipe in 1928, a significantly lower toll than after Category 3 San Ciriaco in 1899, which killed about 3,000 people. Georges in 1998, was a puny Category 1, but one whose rains lashed the entire main island, making it one of the most punishing disasters in Puerto Rico's history. As we know from documented accounts of Maria, there can also be loss of life after the hurricane passes. As Hurricane Fiona also demonstrated in 2022, even a Category 1 hurricane can destabilize the electric grid for weeks, leaving sick, elderly, and disabled people particularly vulnerable.

We follow Steward B. Schwartz in regarding hurricanes as a metanarrative, a widespread organizing theme of Caribbean cultures; hurricanes are pivotal for analyzing Caribbean cultures along with other metanarratives like: "slavery, war, plantations, migration, and colonialism" (2015, xi). Beyond the effects and aftereffects on humans, hurricanes alter the landscape and coastlines and their explosive power vastly exceeds human agency. NASA says they expend as much energy as 10,000 nuclear bombs. [8] Our lack of even a semblance of control over their paths confirms human limitations. Still, people dream of controlling them. One doesn't have to be a climate scientist to know that nuking a hurricane, as Donald Trump was rumored to have suggested in 2019, is a hare-brained idea because it would only spread nuclear fallout without stopping the hurricane. Scientists, however, have considered ways to mitigate these monsters; NOAA documents how Project STORMFURY, an experimental program from 1962 to 1983, proposed seeding the hurricane eyewall with silver iodide to control its forces. Climate scientists have considered other ideas, but none worked because "a hurricane's power eclipses our engineering schemes" (Waymer 2019).

There are social wounds in the best Puerto Rican narratives and images when they reflect the trauma of disasters. Trauma is another way to study disasters and culture. Peer Ilner and Isak Winsel Holm state that a fundamental point about trauma is that, for Freud, it appears as a trace, "manifesting itself in dreams, anxieties and breakdowns later in life" (2016, 53). This metaphorical feature of trauma resonates in literature and the arts, which excel at exploring hidden meanings through symbolic processes. Cathy Caruth says it best when she states that trauma is "is always the story of a wound that cries out, that addresses us in the attempt to tell us of a reality or truth that is not otherwise available" (1996 4).

Ecocriticism privileges geologic eras that mark climate change, which is why here we use the terms Anthropocene, Capitalocene, and Androcene. Anthropocene considers the historical impacts of humans on the environment. It is not an accepted scientific term because geology's governing body, the International Union of Geological Sciences, decrees that as a geologic designation it is premature and lacks a clear starting point. [9] Its start has been variously pinpointed as: in the Neolithic, or around 1500, or the Industrial Revolution (1800s), or the testing of the first atomic bomb (1945). The Holocene is instead the accurate geologic term for our era, the latest part of Earth's Quaternary era (roughly 2.6 million years). Despite the geologists' decree, we culture scholars, as well as social scientists, politicians, and economists regularly use Anthropocene to label humans as a corrosive geological superpower (Ellis 2016, 192) or a destructive agent "akin to meteorites and volcanoes" (Vermeulen 2020, 4).

The term Capitalocene assigns culpability for the Earth's climate changes, because not all humans share equal blame for planetary environmental degradations. It signals that environmental crises are caused "not by humans in some undifferentiated and generalized way, but more specifically by the global spread of capitalism and its socio-economic-ecological injustices" (Arons 2020, 16). Historian Jason W. Moore, who coined the term Capitalocene, stresses that culpability is systemic: "the phrase *anthropogenic*[10] *climate change* is a special brand of blaming the victims of exploitation, violence, and poverty. A more nearly accurate alternative? Ours is an era of *capitalogenic* climate crisis."

Puerto Rico is the poster child of the Capitalocene because it has suffered colonial regimes of two empires, Spain and the United States, which contributed greatly to a capitalist industrialized world. Western industrialization brought environmental degradation and increased the greenhouse gases responsible for climate change. Land-grabbing from and oppression of indigenous populations as well as the promotion of transatlantic slavery both accelerated the deleterious effects of those powerful overlords as their extraction of minerals and resources fueled the human and environmental ills of industrialization.

To the terms Anthropocene and Capitalocene, we add Christopher Abram's term Androcene: "Although individual women may have contributed in many ways to humanity's detrimental impact on the Earth, they have overwhelmingly done so in patriarchal contexts... the end of the world is the work of men and [Androcene] signals our rejection... of the homogeneous, heroicized, patriarchal 'antropos'" (2019, 166). The term Androcene sharpens this book's analyses on gender (Fig. 1.1).

The most important Puerto Rican artist of the nineteenth century, Francisco Oller y Cestero, painted the tranquil landscape *Hacienda Aurora* in 1898–99, during a meteorological lull between hurricane San Roque in 1893 and the catastrophic San Ciriaco in 1899. Puerto Rico became a U.S. colony as he was painting and, furthermore, he and his family sought refuge from that invasion in the town of Carolina where the sugarcane mill Hacienda Aurora was. The tension in this painting is that, while it embodies specific national socio-political realities, it elides hurricanes.

As this book moves from Oller's foundational landscapes to contemporary art, it grapples with the evident crises of the Capitalocene that impact landscapes. The effects of our contemporary period are detrimental to both the human and the non-human world. Meteorologist and

Fig. 1.1 Francisco Oller y Cestero, *Hacienda Aurora*, 1898, óleo sobre panel. Museo de Arte de Ponce. The Luis A. Ferré Foundation, Inc

atmospheric chemist Paul J. Crutzen's litany of climate change woes is both shocking and exhausting (2002, 23). [11]Pieter Vermeulen rightfully notes that Crutzen's listing of dire facts taxes one's capacity to process information (2020, 4).

This is where the strengths of the humanities help: storytelling and figurative tools such as metaphors, symbols and images that may awaken people to the consequences of climate change. The environmental humanities stress that cultural works can make visible the threats of environmental degradation as well as support political and economic solutions to those problems. Since climate change does not affect everyone equally, artists and writers illuminate systemic inequalities in the degradation of the environment, and may offer ways to restore it. Cultural products may highlight the urgent need to mitigate the Capitalocene's effects through environmental justice, especially for populations that suffer from patterns of vulnerability (Oliver-Smith and Hoffman 2002, 3). The risks are critical given scientific proof that cyclones will worsen, driven by the rise in air and ocean water temperatures. Environmental humanist Christina Gerhardt summarizes the grim prospect of sea levels rising for Puerto Rico: "Puerto Rico would lose 3.6% of its coastal land. The higher water level would flood coastal infrastructure, such as water, pumping stations and pipelines, a waste water treatment plant, and six power plants. It would also inundate the airport in San Juan" (2023, 250).

Puerto Rican culture shows two interwoven tendencies: a deep engagement with the environment and also ecoerasure. This book traces this push–pull pattern throughout centuries of Puerto Rican culture. For example, José Luis González, one of the most prolific twentieth-century Puerto Rican intellectuals, addresses human and environmental crises together in his canonical story "En el fondo del caño hay un negrito" ("At the Bottom of the Canal There Is a Little Black Boy"). For him the poor are as much victims as is the *caño*, both asphyxiated by colonial Puerto Rican modernity. The story famously depicts the racialized socioeconomic disadvantages of the child Melodía's family within their specific location of a *caño*. González grappled with the effects of the Capitalocene in 1950, before it even had a name, as the story shows the privileged riding buses or cars, looking down on the family from the infrastructure of roads that choke the environment. We welcome environmental justice: in 2016 the Caño Martín Peña Community Land Trust reclaimed the "dignity" of the *caño* restoring it for the eight communities around it since the 1920s. [12] However, even though González highlights the environment in his story, it plays no part in his influential essay on Puerto Rican culture, *El país de* cuatro pisos ("The Four-Storied Country").

Roadmap to an *Ecocritical Study of Puerto Rican Culture: Disaster Nation*

Chapter 2. Ecophobia and Ecophilia

Early Puerto Rican culture displays the push and pull of ecophobia (hatred of Nature's violent power) and ecophilia (belief in Nature's benevolence)—as well as the beginnings of ecoerasure ("forgetting" environmental disasters). We explore how Tainos experienced hurricanes using paleotempestology and looking at their complex mythology (with multiple hurricane-causing deities). Traditional patriarchal norms of storytelling, deeply allied to the growth of capitalism, produce stories of manly Androcene heroes facing disaster. A chaotic, unknowable climate fueled heroic tales, a cornerstone of disaster writing. Pirate stories also highlight independent men conquering the environment. Conversely, we examine themes of ecoerasure and women. Religion shaped the understanding of disasters, as disaster tales show, these in turn deepened the colony's ties to the Spanish Crown. Denunciations of governmental disaster inaction also become commonplace, linking hurricanes and colonialism. Chroniclers viewed Nature as a web of timeless forces and believed in natural succession: belief in a resilient Nature that renews after a hurricane. We feature historical writing from the Spanish colonial era up to mid-nineteenth century: Christopher Columbus, Juan Ponce de León, Juan Ponce de León (II or Troche), Gonzalo Fernández de Oviedo y Valdés, Bartolomé de Las Casas, Fray Iñigo Abbad y Lasierra, Alejandro Tapia y Rivera, and others.

Chapter 3. Nation-Building Ecoerasures of Disaster

Natural disasters and the governmental inefficiencies that heighten them are wholly underrepresented in the nineteenth and early twentieth centuries. The unfulfilled dream of Puerto Rican independence bolstered fictionally placid depictions of Puerto Rico's environment. We use Michel Foucault's designation of the garden as a "heterotopia" to explore this aestheticized landscape. We examine early anthologies, Alejandro Tapia y Rivera, Eugenio María de Hostos, José Gautier Benítez, Manuel Zeno Gandía, Lola Rodríguez de Tió, and others. Painters from rococo José Campeche onward also linked national identity to a landscape portrayed as serene. Impressionist Francisco Oller's landscapes are foundational because of his representations of native flora—that which beautifies and feeds the Nation—within a landscape free of natural disasters. We compare Puerto Rican art to global nineteenth-century artists who tackled disaster (Hokusai, Homer). Salvador Brau typifies ecoerasing historical writing. Ecological and environmental representations connect to a contentious debate: docility as a national characteristic. Manuel Zeno Gandía's *La charca* typifies the push–pull forces at work in cultural depictions of the environment. Government and commerce collude with the fiction of *La Isla del Encanto*.

Chapter 4. Climate Anxiety, Trauma, and Infrastructure in the Twentieth Century

By at least mid-century, writers and artists are engaging with facets of ecology, geology, and meteorology in ways that reflect material reality while stressing the dream of political independence. The strength and resilience of Puerto Ricans overcoming disaster is a driver of more contemporary culture. DIVEDCO, a government agency, promoted hurricane preparedness thus tying the arts to disaster. Infrastructure and climate anxiety are constant motifs. Creators denounce toxic inefficiencies, governmental grift and corruption coupled under exploitative U.S. support. We explore writers like Antonio S. Pedreira, Margot Arce de Vázquez, Rosario Ferré, Edgardo Rodríguez-Juliá, Magali García Ramis, Ana Lydia Vega, Eduardo Lalo and artists Eduardo Vera Cortés, José M. Figueroa, Carlos Raquel Rivera, Rafael Tufiño, Lorenzo Homar, and others who produce art that showcases hurricanes.

Chapter 5. Post-Maria Representations of Disaster, Testimony, Decolonizing Resilience and Reconstruction

2017 was a crucial year for Puerto Rican culture when hurricane Maria walloped the island as a Category 4 hurricane. The earthquakes of 2020 also brought unpredictable geological disasters that churned up the culture. We explore precarious living in a violent bioregion with inadequate infrastructure yet where community resilience shines as activists, artivists and climate justice advocates encourage decolonial actions for the environment. Anthologies of testimonials include themes of memory and grief. Migration is also as a force as potent as any disaster. We analyze literary works from Puerto Rican and diasporic authors: Edgardo Miranda-Rodríguez, Ana Lydia Vega, Mayra Santos-Febres, Eduardo Lalo, Tere Dávila, Ramos Isabel and Kisha Tikina Burgos, Ana María Fuster Lavín, Rafael Rodríguez Cruz, Xavier Navarro Aquino, Esmeralda Santiago, E.G. Condé, and others. Art brings protagonism for the non-human alongside humans. We consider themes of ecology and landscape, infrastructure, memory and grief, resistance, and protest as we explore Daniel Lind-Ramos and other artists and we analyze extensively the exhibit *no existe un mundo poshuracán: Puerto Rican Art in the Wake of Hurricane Maria*.

Chapter 6. Conclusion

Puerto Rico is a unique example of both perilous location and lasting colonialism, and a good testing ground for our ecocritical analysis of how cultural attitudes, images and stories affect our views of resilience and preparation for disaster under the effects of the Capitalocene.

Appendix. Catalogue of Disasters and Major Works of Puerto Rican Culture

N.B. All translations in this book are our own unless otherwise specified.

Notes

1. Puerto Rico confronts increases in "sensible and latent heat energy, increasing the thermodynamic potential wind intensity" that will fuel even more disastrous hurricanes (Wehner and Kossin 2024).
2. Wehner, Michael F. and James P. Kossin discuss the need for expanding the scale to number 6.
3. See Danielle Zoe Rivera (2022).
4. Our Appendix documents meteorological and geologic disasters specifically for Puerto Rico.
5. https://web.archive.org/web/20230130071654/https://www.ncdc.noaa.gov/stormevents/eventdetails.jsp?id=5423520
6. Use of gender to analyze humans and the natural world.
7. In Chapter 4 we explore how Magali García Ramis's *Felices días, tío Sergio* reflects the specific trajectory of a storm because its setting is San Juan, which that specific storm bypassed, altering the trajectory of the story.
8. https://web.archive.org/web/20250210155313/https://earthobservatory.nasa.gov/favicon.ico
9. The geological authorities "weren't ready to ratify an epoch that is still so young, at least by the standards of Earth's 4.6-billion-year history" (Zhong 2024).
10. Chiefly environmental pollution from human activity.
11. "Tropical rainforests disappear at a fast pace, releasing carbon dioxide and strongly increasing species extinction. Dam building and river diversion have become commonplace. More than half of all accessible fresh water is used by mankind. Fisheries remove more than 25% of the primary production in upwelling ocean regions and 35% in the temperate continental shelf. Energy use has grown 16-fold during the twentieth century, causing 160 million tonnes of atmospheric sulphur dioxide emissions per year, more than twice the sum of its natural emissions. More nitrogen fertilizer is applied in agriculture than is fixed naturally in all terrestrial ecosystems; nitric oxide production by the burning of fossil fuel and biomass

also overrides natural emissions. Fossil-fuel burning and agriculture have caused substantial increases in the concentrations of 'greenhouse' gases—carbon dioxide by 30% and methane by more than 100%—reaching their highest levels over the past 400 millennia" (2002, 23).

12. Ojeda Matos explains families were driven from the countryside to the *caño* by economic need; in 2022 there were 27,000 people in eight communities.

CHAPTER 2

Ecophobia and Ecophilia

Traditional patriarchal norms of storytelling, deeply connected to the growth of the Androcene's capitalism, feature stories of manly heroes facing disaster and highlight religious beliefs and practices that sought ways to mitigate the abrupt disasters visited upon the Puerto Rican colony. Androcene tales tie Spanish colonial culture to a religious and political system in which Europeans dreaded the new (to them) disasters (hurricanes). These stories demonstrate shared fears of loss of agency and control over the environment since the whole enterprise of conquest and colonization involves desire to control the environment, including humans and non-humans. We trace the genealogy of heroic agency from the 1500s to the nineteenth century as it responds to opposing attitudes to the environment: ecophobia and ecophilia. Early modern Puerto Rican culture displays the push and pull of ecophobia (hatred of Nature's violence) and ecophilia (belief in Nature's benevolence)–as well as the beginnings of ecoerasure (forgetting natural or human-caused disasters).

The environment was of great importance in the lives and beliefs of the Tainos before Europeans arrived, which we see in their complex mythology; this is why we begin by looking at paleotempestology, scientific theories about hurricanes during Taino times. We then look at narrative practices about the natural world that respond to the socio-political forces of the early Spanish colony, particularly the over-arching

M. Acosta Cruz, *An Ecocritical Study of Puerto Rican Culture*, https://doi.org/10.1007/978-3-032-04044-2_2

religious ethos of those times. We will foreground these tales with scientific and historical data.

The seeds for ecophilic, ecophobic, and (later) ecoerasing views of the landscape were laid in early modern times. This chapter considers how early Puerto Rican culture processed disaster, exploring chroniclers like Cristopher Columbus, Juan Ponce de León, Bartolomé de Las Casas, Fray Iñigo Abbad y Lasierra, Antonio de Herrera y Tordesillas, and Juan de Castellanos. In their writing we see explanations of a violent bioregion when people could not predict hurricanes (as we now can) or earthquakes (which we still cannot). Puerto Rican chroniclers toggle back and forth between views of a disaster-ridden region and its opposite, an Edenic land of self-renewing nature. These opposites will remain in Puerto Rican culture until at least the twentieth century. Framed by accounts of manly Androcene heroes battling a ferocious Nature and devout prayers for protection from hurricanes and earthquakes, many of the early depictions of the new, unspoiled land seesaw between the "pastoral and apocalypse" (Garrard 2012, 2).

Paleotempestology, Climate and Deities Before Written History

We know people inhabited Puerto Rico as early as 4,200 B.C.E. (Schnabel 2023) but, how bad were the storms they lived through? Climate scientists Daidu Fan and Kam-biu Liu define paleotempestology as a branch of historical climatology that, since the early 1990s, has studied pre-Columbian climate: "past typhoon activity spanning several centuries to millennia (usually 6000 years) before the instrumental record era, mainly through the use of geological proxy techniques" (2008, 2908).[1] Gillen D'Arcy Wood, whose research combines English and Earth Science, writes that these climatologists, "have mined a multitude of proxy sources—from ice cores, tree rings, and fossilized pollens to ships' logs and 'weather' diaries—to fill out the climate record (albeit incompletely) across a breathtakingly broad temporal range" (2008, 1–2).

Paleotempestology offers several theories about the long-term fluctuation of hurricanes.[2] Paleotempestologists E. J. Wallace et al. concentrate on the Bahamas and conclude there were four periods of intense hurricane activity over the past 1,500 years (2019). Erick Braun et al. state that hurricane activity in the areas from New York to Puerto Rico was higher between 2,000 and 1,100 years ago (2017, 366, 371). In addition, Frank

Oliva et al. see low activity between 950 AD and 1700 in the U.S. Atlantic coast and the Caribbean with a sudden increase around 1700 (2018). This last intriguing possibility would support the Taino belief (as reported by early historian Fray Iñigo Abbad y Lasierra) that Spaniards brought the cursed weather with them.

We know Tainos processed hurricanes through their complex mythology; the multiplicity of hurricane-causing deities they had tells us they regarded hurricanes as major characters of the environment. The Tainos had two main deities responsible for hurricanes, the goddess Guabancex and the god Juracán, a malignant spirit who, according to critic Luis A. Medina, acted as Guabancex's weapon (N.d.). Art historian Joseph R. Hartman reports Guabancex and Juracán were mother and son (2023, 127). Some regard Juracán as an entity in his own right and historian Fernando Picó explains Juracán's key contribution to Spanish: "había una personalidad furiosa y dañina, Juracán, de cuyo vocablo se derivó la palabra castellana huracán, para denominar las tormentas enormemente destructivas del Caribe" ("there was an angry and baleful personality, Juracán, from which word Spanish derived the word hurricane to designate the enormously destructive storms of the Caribbean"; 1990, 26–27). Tellingly, Picó neglects to include the female deity.

Quoting chronicler Fray Ramón Pané, who lived in the area in the late 1400s, the great woman of letters Josefina Rivera de Álvarez explains the multiple ways Tainos thought about Guabancex, mentioning two helper deities, Guatauba, a herald under her command, and Coatrisquié, who gathers the waters of valley and mountain for her (1983, 5). Guabancex is also a trickster and a rebel: "Guabancex es como la hija mala que desobedece las leyes de madre tierra enviando huracanes" ("Guabancex is like the bad daughter that disobeys the laws of Mother Earth by sending hurricanes" (Santiago Álvarez N.d., 4). In addition, archaeologist Jago Cooper lists hurricane deities, Boinayel and Coatrisque (sic) (2012, 11).

We also know that the Tainos dealt with hurricanes from the symbolism in their artifacts. Their clay pots sometimes portray figures with spiral arms, or centripetal "sigmoides,"[3] representations which become part of the iconography of hurricanes. Fernando Ortíz saw these Taino symbols: "[e]n el museo de la Universidad de Puerto Rico hemos encontrado varios dibujos de simples espirales en las asas de las típicas ollas de barro y en ciertos discos" ("in the museum of the University of Puerto Rico we have found several drawings of simple spirals on the handles of typical clay pots and on certain discs"; 1947, 205). These iconic symbols prove that the Tainos experienced hurricanes on a regular basis. Science backs this theory; a team of paleotempestologists studied parts of Vieques and concluded that, "although the *Laguna Playa Grande* record exhibits large-scale changes in hurricane frequency on centennial to millennial time scales, the magnitude of these events has stayed relatively constant" (Woodruff et al. 2008).

There is a great contrast between what the native inhabitants knew about the environment and what the colonizers faced, since hurricanes do not exist in Europe: "Europeans came to the islands unprepared for the destructive storms, even as indigenous people understood their massive power" (Korten 2019). The conquerors' climate anxiety about these fierce conditions would often turn into ecophobia, fear of the natural environment. Simon C. Estok, who coined the term, says that "[f]ear of the loss of agency and fear of the loss of predictability are what form the core of ecophobia" (2009, 134). Christopher Abram's concept of Androcene (see Chapter 1) fits the narrative of Puerto Rico from its earliest times. Spanish chroniclers often told tales of manly men facing hurricanes whose power to transform the land and alter people's lives also influenced how the Spanish Crown would respond to the needs of the colony.

With their wholly inexorable violence hurricanes bring thrilling and entertaining drama to the chroniclers' stories. Juan Ponce de León, who was appointed governor by the Crown in the early 1500s, wrote the earliest testimonial account of the first recorded hurricane in Puerto Rico in 1508. He set an important narrative template because he is both witness to and manly hero of disaster. Writing in 1509 to the Governor of the Indies, Nicolás de Ovando, Ponce de León portrays himself as the hero of the story, representing himself as a busy traveler loyal to the Crown, one who has battled two hurricanes. Many of the earliest chroniclers wrote public relations tales vying for favor from the Crown, and Ponce de León is nothing if not the hero of a sensational seafaring adventure, praising his own survivalist skills. Storms are fore and aft to his long tale of how he conquered hurricanes, as well as everything else in the Caribbean environment including people, land, and sea. He centers his tale on being the intermediary of the Crown with the Tainos. It is a narrative bookended by hurricanes:

> estando en el puerto de Yuma (en la Española) a tres de agosto (1508), vino tal tormenta que metió el caravelón sobre unas peñas, y de allí lo saqué

> e perdí mucho del bastimento. Item: después de pasada dicha tormenta, me partí siguiendo mi viaje, e fuí a la ysla de mona, donde allé a los caciques e yndios de dicha ysla de sanct xoan por la parte del sur, a doce de agosto de dicho año donde surxi de la playa que está en el paraxe del cacique Agüebana, e fuí a su casa, e ablé de parte de vuestra merced lo que me mandó, e le aseguré, e le mandé facer un conuco para Su Alteza, e él dixo que lo faría; e dempues me han dicho que lo ha fecho en non e podido ser ynformado de qué tamaño es, ni le e podido yr a ver, a cabsa de las muchas ocupaciones que he thenido destar apartado del assiento, que thengo comenzando a facer, segund adelante diré. Estando el dicho caravelón sulro (sic), a diez y seys de agosto vino otra tormenta, que le sacó a la costa, y le saqué con mucho trabaxo. ("While in the port of Yuma (in Hispaniola) on August 3 (1508), such a storm came that grounded the ship on some rocks, and from there I removed it losing many supplies. Item: After that storm passed, I set out on my journey and went to the island of Mona, where I found the caciques and Indians of said island of Sanct Xoan in the southern part, on August 12 of said year where I emerged from the beach that is the seat of the cacique *Agüebana*, and went to his house, and spoke on behalf of your grace what you ordered, and assured him, and ordered him to make a parcel for His Highness, and he said that he would do it; and later they told me that he did and I have not been able to see how big it is, nor have I been able to see it, because with many occupations I have being away from the seat, these I have begun to do, as I will say. While said vessel was loose on August 16th another storm came, which pushed it to the coast, and I recovered it with a lot of effort"; Salivia 1950, 4)

Reflecting the mentality of the times, Ponce de León's tale has specific hierarchical ratings: (1) Heroic Man (in this case himself) battling hurricanes; (2) the colonial pecking order, with due deference to his superiors and the Spanish Crown; (3) a manly life beset by enemies; (4) the importance of Crown-awarded territory; and, of course, (5) the dependence on ships, which could be wrecked by hurricanes. Ponce de León is accounting for his governance of the island while stressing his hurricane-fighting prowess to support his formal status as representative of the Spanish Crown before the Taino *cacique*. By virtue of Ponce de León's high governmental status, he is the center of the narration; he does not, for instance, dwell on his crewmen. His account of these two hurricanes therefore sets a template in which disasters give prominence to heroic Men (with one notable exception we will soon see) battling *mano a mano* against hurricanes. The heroes in early writings were, of course, men of higher social status, which included Taino *cacique* (chieftain) Agüeybaná.

Ponce de León's attitude toward Nature corresponds with Western thinking of the times, especially the idea of a "designated earth" in which the relation between man and nature is one of earth "designed for people alone, as the highest beings in creation, or for the hierarchy of life, with man as the apex. This conception presupposes the earth or certain parts of it to be a fit environment not only for life, but for high civilization" (Grove 1995, 25). The Catholics colonizing the region, however, soon realized that the hierarchy of Being had no effect on hurricanes or earthquakes (which they thought causally tied). Caribbean Nature was unpredictable and ungovernable and put everyone at the mercy of forces the magnitude of which they could not predict nor comprehend.

This loss of control over Nature undergirds the stories early writers tell; lack of knowledge and control are crucial storytelling elements in early modern stories. Anyone who has experienced a hurricane understands Rudolph Adams Van Middledyk[4] when he says, "Whoever has witnessed the awful magnificence of what the primitive inhabitants of the West Indian islands called *ou-ra-cán*, will never forget the sense of his own utter nothingness and absolute helplessness" (1903, 213). The dread of unpredictable hurricanes during the early colony pervades their tales. This level of climate anxiety is true from the early colony to this day, the difference is now we see hurricanes as they form and move from the west of Africa.

Chronicler and human rights activist *avant la lettre* Bartolomé de Las Casas, who arrived on the island in 1502, features hurricanes many times in his *Historia de Las Indias* (1527–1561). He says one storm caused 7,000 indigenous deaths although, alas, he gives no date. He also praises Columbus for knowing when storms were coming—this will be a motif in early histories about disasters although Las Casas also equally praises indigenous knowledge of the environment.

Earthquakes, taken at the time to accompany hurricanes, provide another source of early modern climate anxiety. The first thorough historian of Puerto Rico, Fray Iñigo Abbad y Lasierra, witnessed the hurricane of Aug 28, 1772, stating it also caused an earthquake. The two cataclysms would remain tied in the imagination of many (1788, 431).

Ecophilia, love and knowledge of and connection to the natural environment, coexists with ecophobia in the early historians. The chroniclers framed stories around heroes who overcome disasters to reaffirm views of

Man overcoming the wild cataclysms of this new Caribbean environment. As mentioned in Chapter 1, many chroniclers rate Columbus as the first hero to predict hurricanes because he foretold the storm of 1502 in La Española that caused his enemy Bobadillo's death. Nearly all the chroniclers repeat this story because its inherent drama feeds into the myth of a Great Mariner who knew the environment and foresaw disaster. Las Casas reports the reasons Columbus could see the coming of the storm in *Historia de Las Indias* (1550): "una señal muy eficaz de haber de venir tormenta y que por maravilla yerra, es cuando sobreaguan muchas toninas, que son creo los que llaman por otro nombre delfines, y los lobos marinos" ("a very effective sign of a coming storm and which hardly errs, is when many tuna come above water, which I think are called by another name dolphins, and sea lions"). These observations were astute; today many scientific studies prove there are links between animal behavior and storms: birds use infrasound to sense storms, toads can sense earthquakes, all animals react to barometric pressure and frogs croak louder when a storm is coming.[5]

The myth of Columbus's otherworldly connection to the environment, to seas and islands that he appears to know as well as the Tainos were reported to, burnishes the image of the heroic discoverer and consequently makes him a fitting Androcene hero of early Modern histories. The ecophilic legend of Columbus's knowledge and therefore control of the environment has persisted. Historian Edwin Miner Solá calls him the first meteorologist of the Americas, saying he saw signs of bad weather in the clouds, the conditions of the sea, and the behavior of certain fish (1995, 2). Climatologists Bob Sheets and Jack Williams, in *Hurricane Watch,* their book about the history and science of forecasting hurricanes, say Columbus knew Nature, "the Admiral of the Ocean Sea had learned a great deal about tropical weather during his four journeys of exploration" (2001, 3). They credit him with predicting a coming storm due to his know-how of cirrus cloud movement and dolphin behavior. And yet Columbus also denied the brutal environment when it suited him; his 1493 letter to Santángel portrays the region as a placid paradise of gentle people eager to serve. The same promotional intent is evident in other colonial correspondence.

Antonio de Herrera y Tordesillas, an ecophobic early historian who never set foot in the Americas, was mostly concerned with recounting human affairs, especially heroics. He describes two heroic factions, that of the famous Taino chieftain Agüeybaná and the opposing Castilian conquerors. In 1601–1615 Herrera writes the clearest description of how the colonizers themselves were an ecological disaster for the bioregion, chronicling how Ponce de León and his men came back to San Juan from Florida to abundant quarry:

> En un rato de la noche tomaron, en una de estas islas, ciento y sesenta tortugas, i tomaran muchas más, si quisieran: i también tomaron catorce lobos marinos, y se mataron muchas alcatraces, y otras aves, que llegaron a cinco mil. ("In a little while in the night they took, on one of these islands, one hundred and sixty turtles, and they would have taken many more, if they wanted to: and they also took fourteen sea lions, and they killed many gannets, and other birds, which reached five thousand in number"; 1601, 248)

The sheer number of animals killed and Herrera's ecophobic triumphalist tone shocks contemporary readers for whom this behavior, with its wanton destruction against non-humans, is an ecocide.

Like hurricanes, tough and grasping European men arrived in somebody else's unspoiled land eager to remake their identity into heroes and conquerors by controlling and overcoming everyone and everything. Except, of course, they had no control at all over hurricanes and earthquakes. Poet Juan de Castellanos's *Elegías de varones ilustres de Indias* ("Elegies of Illustrious Men of the Indies," 1589) is another ecophobic book full of heroes. Castellanos arrived in Puerto Rico in the 1540s and his epic poem celebrates heroic colonizers triumphing over both native people and the environment. In a gory battle scene between Spanish "fuertes capitanes" ("strong captains") and Tainos, Castellanos describes the disaster the Europeans represented for the Caribbean using "hurricane" as a metaphor for their destructiveness. He celebrates the killing of indigenous people using the exact traits of hurricanes to show the treatment the Spaniards meted to the Tainos:

> Sus fuertes capitanes animando / Que muchos le vinieron con presteza / Sangre, cascos y sesos derramando / con una nunca vista lijereza . . . / Bien como *huracán*, que de tal priesa / en indicas provincias y regiones, / que barre la montaña más espesa, / quebranta ramas, vuelve los troncones, / y los anchos caminos atraviesa / con crecidísimas inundaciones, / causando tal temor a los humanos, / que quedan como muertos los más sanos / . ("Cheering their strong captains / many came with alacrity / Blood, helmets and brains spilling / with a unprecedented nimbleness . . . / Like

> a *hurricane*, that in such a hurry / in Indie provinces and regions, / sweeps away the thickest mountain, / breaks branches, overturns trunks, / and crosses the wide roads / with massive floods, / causing such fear to humans, / so that the healthiest are as dead"; 1847, 118, our emphasis)

Castellanos vividly constructs his then-innovative simile of the (for him righteous) hurricane-like violence of the heroic *varones ilustres* with specific details of real cyclonic activity: squalls, uprooted vegetation and floods.

Hurricanes often take center stage in these early histories. Gonzalo Fernández de Oviedo Valdés, a botanist and one of the very first chroniclers of the Americas, wrote the first comprehensive history of Spanish America, *Historia natural y general de las Indias* ("Natural and General History of the Indies"), recounting events from 1492 to 1549. He says: "[h]uracán, en lengua desta isla, quiere decir propriamente tormenta o tempestad muy excesiva" ("hurricane, in the language of this island, properly means storm or very excessive storm" 1535). Oviedo wrote five years after 1530, a disastrous year when three hurricanes walloped Puerto Rico.

San Juan-born Juan Ponce de León (II or Troche) and Antonio de Santa Clara's 1582 *Memorias de Melgarejo* praises the colony and describes disasters because 1568 saw hurricane San Bartolomé and 1575 hurricane San Mateo.[6] *Memorias* praises the island's environment, from its waters to its guava trees, saying the "temperament" of the island is "good." Yet Troche adds it is so except in August and September when it is visited by *juracanes* which "las veces suelen hazer grandísimos daños bentando los vientos" ("usually cause great damage blowing their winds"; 1582, 112). *Memorias* states hurricanes happen every ten or so years. It is a very early work of ecological history and classifies Puerto Rico's botanical bounty, including an extended description of a monumental *ceiba* tree that stretches across a river. The *ceiba* will become a beloved national symbol for Puerto Rico's resilient Nature. A centenary *ceiba* with deep roots and wide canopy is a wonderful symbolic anchor for love of Land. Later Enrique Laguerre's novel *La ceiba en el tiesto* ("The *Ceiba* in the Flowerpot"; 1956) presents the *ceiba* as a powerful metaphor against the denaturalized constraints of colonialism. Artists like Francisco Oller, Rafael Tufiño, and Antonio Martorell also feature the *ceiba*, a powerful landmark that withstands most storms.

One of the most famous historians of his time, Fray Iñigo Abbad y Lasierra lived in Puerto Rico for ten years while he was secretary to the bishop. A careful observer of hurricanes and earthquakes he links the two, calling them fearsome and costly (1788, 211). He witnessed the 1772 hurricane, saying it also featured "un temblor lento de la tierra" ("a slow earth tremor") along with winds, rains, thunder, and lightning (1788, 431). He paints a vivid picture of the impact of a hurricane:

> El huracán es el fenómeno mas horroroso de cuantos se observan en esta isla y aun creo que en toda la América. Es un viento furioso acompañado de lluvia, relámpagos, truenos y las más veces de temblores de tierra, circunstancias todas las mas terribles y devastadoras que pueden unirse para arruinar un país en pocas horas: los torbellinos que forma el aire y las torrentes que inundan los pueblos y campiñas, acompañados de relámpagos, parecen anuncian las últimas convulsiones del universe. ("The hurricane is the most horrifying phenomenon of those observed on this island and I even believe in all of America. It is a furious wind accompanied by rain, lightning, thunder and most often by earthquakes, the most terrible and devastating circumstances that can ruin a country in a few hours: whirlwinds that form in the air and torrents that flood the towns and countryside, accompanied by lightning, seem to announce the last convulsions of the universe"; 1788, 120)

Like other early modern historians he has an apocalyptic tone, making clear the superlative power of that clearly traumatic event. He recounts at length stories of earlier hurricanes; for instance, he reports there were two hurricanes in 1530[7] that left devastation in their wake, widespread famine and poverty and, even worse, the Caribs attacked since they too were desperate for food.

Remarkably, Abbad tells of a woman hero during one of those early hurricanes. Women rarely figure or are mentioned in chroniclers' worldview, but even the so-called weaker sex could find agency in the aftermath of apocalyptic storms. Abbad tells the following rescue tale after the 1530 storms and the subsequent Carib invasion:

> La muger de Cristóbal de Guzmán que era rica y amaba á su marido, viendo que de Santo Domingo no veían fuerzas suficientes para castigar a los Caribes, resolvió armar cinco vergantines esperando rescatarlo [los rescatadores] desembarcaron de noche y vinieron sobre los Caribes con tanta rabia que hicieron en ellos gran mortandad, cautivaron muchos y recogieron los negros, indios y españoles que se habían llevado de Puerto Rico. ("Cristóbal de Guzmán's wife, who was rich and loved her husband,

> seeing that not enough forces came from Santo Domingo to punish the Caribs, decided to arm five ships hoping to rescue him [the rescuers] disembarked at night and fell on the Caribs with such rage that they greatly slaughtered them, captured many, and retrieved the Blacks, Indians and Spaniards who had been taken from Puerto Rico"; 1788, 121)

This nameless wife's disaster story fits the European mentality of conquerors who regarded indigenous people as enemies to be killed. Her heroic tale, moreover, showcases a woman's sense of community, since the mix of people who were saved consists of indigenous people, Blacks and Spaniards. Even more exciting than the heroine's doomed rescue attempt of a husband (alas, already dead), is Abbad's story of a storm that is itself the hero in an ecophilic story in which Nature is an ally of the Spanish colony. This happens when a hurricane saves the city of San Juan from an English invasion:

> En 1678 el Conde de Estren, comandante inglés, armó 22 bajeles y con un cuerpo de tropas de desembarco se presentó a la vista de la ciudad amenazando su destrucción si no se rendía; . . . antes que le dejasen pisar la tierra sobrevino un huracán tan violento que dio con toda la escuadra inglesa sobre la isleta Aves, en donde se hizo pedazos con la mayor parte de las tripulaciones y tropa, quedando prisioneros de guerra los que se salvaron del naufragio. ("In 1678 the Count of Estren, an English commander, armed 22 vessels and with a body of landing troops appeared in front of the city threatening destruction if it did not surrender; . . . before they could set foot on land, there came a hurricane so violent that it threw the entire English squadron to the islet of Aves, where it was smashed to bits with most of its crews and troops, those who were saved from the shipwreck became prisoners of war"; 1788, 184)

On this rare occasion, a disaster brings good fortune as the wrecking power of the hurricane saves the "nation" (before there was any such thing). Yet most of the time hurricanes were the opposite of Abbad's ecophilic and triumphalist tale, bringing only ecophobic misfortune. Historian Manuel Ubeda y Delgado reports that in 1708 locals formed a squadron of ships to attack English invaders, but a storm defeated them and decimated the squadron; all ships foundered killing everyone on board (1878, 22).

Even though present-day technological and mathematical forecasting of hurricanes is infinitely superior to anything in the Taino era, we are closer to their awareness of natural disasters than we are to the cluelessness of the Spanish conquistadors. Abbad views the Tainos's relation to Nature as that of magical natives: not noble savages, but a knowledgeable people. The unknowability of Caribbean hurricane formation, strength, and trajectories caused climate terror in Europeans. Abbad says the opposite was the Tainos's deep connection with the environment, saying that their fine-tuned understanding of their natural surroundings gave them two to three days hurricane warning:

> Los indios de esta isla preveían esta infeliz catástrofe y la tenían por cierta, cuando observaban el aire turbado, el sol rojo, un ruido sordo subterráneo, el círculo de las estrellas obscurecido con un vapor que las aparentaba más grandes, los horizontes por el noroeste cerrados, un olor fuerte que exhalaba el mar, el levantarse este en medio de la calma, de repente de este a oeste. La experiencia de estos temibles sucesos les había enseñado a observar las mutaciones de los astros y elementos y a pronosticar en ellos tan fatales fenómenos. Hoy mismo los anuncian dos o tres días antes que sucedan. ("The indians of this island foresaw this ill-fated catastrophe and took it for a certainty when they observed the troubled air, the red sun, a subterranean rumble, the circle of stars obscured with a vapor that made them appear larger, the horizons to the northwest closed-in, a strong smell from the sea rising in the east in the middle of the calm, suddenly from east to west. The experience of these fearsome events had taught them to observe the mutations of the stars and elements and to predict such fatal phenomena. Nowadays they announce them two or three days before they happen"; 1788, 430)

Abbad's tale engages all the senses as he describes the Tainos' understanding of a totality of their environment, even the stars. He states the indigenous people were finely attuned to the minutiae of the Caribbean environment, including land, sea, and air.

Contemporary archeology assesses Taino hurricane-coping skills positively. Archaeologist Jago Cooper says the construction of their houses[8] shows preparedness for wind shear, coastal storm surges, and pluvial flooding:

> [they] had management strategies in place to identify oncoming hurricanes, alert the community, and retreat to nearby caves for shelter during a hurricane's short-term impact, then their Pre-Columbian house structures were extremely resilient to the mid- to long-term impacts because they could be

> rebuilt within days of a hurricane using locally available and easily sourced materials. (2012, 13)

Disaster heroes change markedly from early modern to more modern times as everyday folk assume heroic roles in battling hurricanes. For instance, Rafael W. Ramírez de Arellano's 1932 *Los huracanes de Puerto Rico* ("The Hurricanes of Puerto Rico") tells stories of two nineteenth-century working-class heroes, to whose adventures he devotes full passages. This history also highlights the major societal problem of that era: slavery. The first tale comes from the devastating aftermath of San Narciso in 1867, one of the top ten hurricanes until 1950. This section starts with local flavor mentioning the hero's town:

> El [héroe] de Carolina es como sigue: . . . del heroico comportamiento del esclavo llamado Alejo Aruz, que, en los momentos de mayor conflicto, cuando el huracán arrancaba las casas, cuando todos procuraban salvar sus vidas, se traslada infatigable, a todos los lugares donde escuchaba el clamor de las mujeres que imploraban misericordia rodeadas de sus pequeñuelos. . . . Alejo Aruz resistía como un hércules el empuje de los vientos. Sus propósitos de salvar a muchas personas llevándolas sobre sus hombros, no pudieron ser más dignos de alabanza. Y a no haber sido por la actividad incansable de Aruz, verdadera providencia para aquel pueblo durante las horas de mayor tribulación, las madres hubieran visto desaparecer sus hijos arrebatados por el furor de la tempestad. Este esclavo es digno no sólo de su libertad sino de la estimación que procede de grandes virtudes y de grandes sacrificios. ("Carolina's [hero] is as follows: . . . the heroic behavior of the slave Alejo Aruz, who, in moments of greatest conflict, when the hurricane ripped away houses, when everyone was trying to save their lives, tirelessly goes to all the places where he heard women imploring for mercy surrounded by their little ones. . . . Alejo Aruz resisted the force of the winds like a Hercules. His resolve to save many people by carrying them on his shoulders could not be more praiseworthy. And had it not been for Aruz's tireless activity, a true providence for that town during its hours of greatest tribulation, mothers would have seen children disappear carried away by the fury of the storm. This slave is worthy not only of his freedom but of the esteem that comes from great virtues and great sacrifices"; Salivia 1950, 201)

The story elevates Aruz to an Androcene hero battling that ultimate antagonist, the hurricane. Notable because of the hero's civic devotion to others, it is not the authorities that save the townsfolk, instead a marginalized figure helps them, not for glory nor to ingratiate himself to the hierarchy (like Columbus or Ponce de León). Aruz rescues those regarded as socially weaker, children and women, specifically mothers, a key building block of a traditionally patriarchal society. Ramírez de Arellano tells of a second hero fighting the hurricane, this one in the town of Gurabo:

> Cuando la tempestad rugía amenazando destruirlo todo, en aquellas horas de mayor conflicto, un esclavo rompe las puertas del calabozo de la ciudad donde estaba preso y encerrado, y obtenida su libertad, lejos de emprender la fuga, para no ser habido, colocase al lado de la autoridad, y salva de una muerte trágica y segura a algunas personas, conduciendo unas al edificio de la alcaldía y otras al mismo calabozo de donde poco a él mismo había salido. . . . Llamábase este último esclavo Juan Quiñones. ("When the storm raged threatening to destroy everything, in those hours of greatest conflict, a slave broke the doors of the city dungeon where he was imprisoned and, obtaining his freedom, far from escaping in order to not be captured, he stands on the side of authority and saves people from a tragic and certain death, leading some to the mayor's office and others to the same dungeon from which he had escaped. . . . This slave was Juan Quiñones"; Salivia 1950, 201–202)

This thriller of a story features a raging storm, a hero who breaks out of jail but instead of fleeing saves others from certain death, and at the very end the reveal that it was a slave who saved them!

Religious Responses to Hurricanes

In Spanish as in English the word "disaster" comes from the French or Provençal *désastre* from *dis* meaning separation, plus *astro* or star, therefore, it means "ill-starred." In other words, a calamity that occurs because the stars (i.e., unknowable, distant forces) send destruction from the world-beyond-us. How did people cope with such misfortune in early times? Twentieth-century historian José Luis Vivas gives context to the infrastructure and living conditions almost unimaginable to us when he describes hurricanes in the early 1500s, showing how vulnerable the area was to the onslaught of tempests and comparing it with conditions in 1956:

> un total de cinco [huracanes] azotaron la isla durante los años 1530 al 1539. Si pensamos en la situación de los colonos, en la carestía de casas de mampostería y en el apartado de los establecimientos, podemos comprender el atraso que representó para la colonia el azote sucesivo de estas tormentas tropicales. Verdaderamente era poco halagüeño para posibles colonizadores el venir a una isla que en el corto lapso de siete años sufre cinco tormentas devastadoras. Recordemos el estado de la isla después del paso por ella del huracán Santa Clara, en una época en que contábamos con radar, medios de aviso, construcciones de hormigón, y podemos tener idea del estado de la isla en aquella época después de tres tormentas en un mismo año: 1530. ("a total of five [hurricanes] struck the island from 1530 to 1539. If we think about the settlers' situation, the scarcity of masonry houses and the distances between structures, we can understand the setback that the successive scourge of these tropical storms represented for the colony. It was truly unappealing for potential settlers to come to an island which in the short span of seven years suffered five devastating storms. Remember the state of the island after the passage of hurricane Santa Clara, at a time when we had radar, methods for warning, cement constructions, and we can have an idea of the state of the island at that time after three storms in the same year: 1530"; 1962, 88)

Furthermore, we know that the three hurricanes of 1530 were preceded by hurricanes in 1508, 1514, 1515, 1516, and 1526. A designated hurricane season, starting from late summer (sometimes June) until end of fall, was the only form of prognostication until 1837 when the barometer was introduced.

Lacking reliable forecasting, one way early modern people coped with these disasters was to trust their Catholic faith because religion, popular beliefs, and superstitions gave a semblance of control over disasters. Prayers and the belief in miracles fuel the thinking that natural disasters were divine punishment for collective sins. Columbus's son Hernando Colón reports in *Historia del Almirante* ("History of the Admiral"), that in 1502 his father used prayer to repel a "manga" or waterspout. Cultural historian Teodoro Vidal reproduces this tale with an illustration of the incident showing Columbus, prayer book in hand, aiming his sword to "cut" a waterspout threatening his ship (2010, 241–242).

Catholic theology processed natural phenomena through its belief system, thinking that supernatural forces, God or Satan, brought disaster causing extended destruction and pain; hence storm activity in the early colony triggered faith-based explanations for natural catastrophes. Las

Casas, writing between 1527 and 1561, says that hurricanes had been a rarity in the area before Spain's arrival and that their increase signaled a punishment from God. An early example of ecoerasure, he declares that in the early days of the colony storms were both infrequent and not life-threatening (this last surely poetic license, considering the three storms of 1530) but then they became more frequent and destructive because of people's sinful behavior:

> [los huracanes] eran muy raros, que por maravilla solían venir sino de muchos á muchos años, como tenemos los viejos en estas tierras experiencia, y, por tanto, poca diversidad y variedad ó transmutación en el tiempo, y por consiguiente, en las personas causaban; haberlos de pocos años acá cada año, y hacer los estragos y destruiciones que por la mar y por la tierra han hecho hacer, otra causa oculta que la naturaleza debemos de buscar, muy nuevos y no es otra cosa que *por nuestros muy nuevos y muchos pecados*. ("[hurricanes] were very rare, which as a wonder used to come only between very many, many years, as we old people have experienced in these lands, and therefore caused little change and variability, or transmutation over time, and having them here every few years, causing the havoc and destruction that they have caused by sea and land, another hidden cause the nature of which we must seek is very new, and none other than *for our very new and many sins*"; 1527–1561, 95, our emphasis)

Another early form of ecoerasure appears in Gonzalo Fernández de Oviedo y Valdés's belief that the conquest itself was beneficent as he asserts that storms stopped after the Spaniards arrived. For him the devout Spanish intervention in the so-called New World worked magic against disasters. He writes that events in Hispaniola in 1509—just one year after Ponce de León suffered the storms of 1508—were marked by the (entirely false) absence of storms:

> después que el Santísimo Sacramento se ha puesto en las iglesias e monasterios desta cibdad, e de las otras villas desta isla, han cesado estos huracanes, desto ninguno se debe de maravillar, porque, perdiendo el señorío desta tierra el diablo, e tomándola Dios para sí, permitiendo que su sagrada fe e religión cristiana en ella sea plantada e permanezca, diferencia ha de haber en los tiempos, e en las tempestades e tormentas . . . después acá, cesaron estos peligros y espantables huracanes o tempestades. ("after the Blessed Sacrament was placed in the churches and monasteries of this city, and in other towns of this island, these hurricanes have ceased, no one should be astonished, because the devil losing the lordship of this land and God

> taking it for himself, allowed his sacred faith and Christian religion to be rooted and remain, there must be a difference in the weather, and in tempests and storms . . . since then these dangers and frightful hurricanes or storms stopped" 1535)

Despite all the pious tokens of Catholic faith, reality continually challenged the colony with hurricanes and earthquakes. Oviedo knew this since he wrote in 1535, five years after the triple disaster of 1530.

The early modern authorities themselves turned to ecoerasure in their eagerness to dismiss storms as a thing of the past. Oviedo, who is typical of this kind of magical thinking, proclaimed storms were over after Spain arrived. He quotes "Instrucciones" from the King of Spain in 1511 to the Mayor and Bailiff of the island: "ha placido a nuestro señor de cesar en ellas [Puerto Rico y La Española] las tormentas y terremotos que antes venían de las cuales sucedían los daños que sabéis y habéis visto" ("It has pleased Our Lord to stop in them [Puerto Rico and Hispaniola] the storms and earthquakes that used to come which caused the damage that you know and have seen") (Salivia 1950, 387). Despite this fake news, disasters soon became normalized as part of the environmental viewpoint of early historical writing in the colony.

Oviedo's stories are frequently contradictory—a trait most early colonial historians share—since he also devotes many pages to tales of storms. His motivations are mixed, first there are storytelling demands, since disasters yield dramatic stories. Second, it was also common at the time to stress the dangers from storms to get the Crown's attention in order to receive subsidies, storm aid, and tax breaks. Oviedo says that Satan could cause destruction on a fearsome scale:

> quien hobiera visto e pasado algún boscaje de grandes y espesos árboles, donde haya acaescido algún huracán, habrá visto cosa de mucha admiración e grima espantosa. Porque están innumerables e poderosos árboles arrancados, e las raíces dellos tan altas, cuanto tovieron lo más encumbrado de las ramas algunos dellos; otros quebrados por medio y en partes, e desgajados e hendidos de alto a bajo; otros están puestos sobre otros de tal manera, que paresce luego ser obra diabólica. ("whoever has seen and passed by some forest of great and thick trees, where a hurricane has occurred, will have seen something to cause much astonishment and frightful disgust. For there are innumerable and mighty trees uprooted, and the roots of them so high, as they touched the highest of the branches some of them; others broken in half and in parts, and torn and cleft from

> high to low; some are positioned over others in such a way that it then seems to be diabolical works" 1535)

Oviedo describes uprooted trees, a destruction he believed the devil had caused: "paresce, como he dicho, ser artificio e obra en que ha entendido el diablo o parte de la comunidad del infierno, e no hay ojos humanos de cristiano que sin espanto lo puedan ver" ("it seems, as I said, to be artifice and work in which the devil or part of the community of hell has been involved, and there are no Christian human eyes that can see it without horror" 1535).

Abbad also speaks of the need for prayers. First, he sets the scene with the pre-hurricane signs from Nature; then he describes how the community tries to forestall a storm through extended prayers:

> Dos días antes de suceder este huracán el cielo estaba enteramente cubierto: la resaca y demás señales ya insinuadas vistas en esta ocasion persuadieron a los moradores de aquel partido la proximidad de este terrible azote, recurrieron a implorar la misericordia divina con dos días de rogativas públicas en la ermita de Nuestra Sra. del Espinal, pero no fueron oídos y al tercer día experimentaron esta desgracia y con ella el hambre, miseria, enfermedades y muertes. ("Two days before this hurricane happened, the sky was completely overcast: the riptides and other signs already as seen on this occasion persuaded the residents of that district of the proximity of this terrible scourge, they resorted to imploring divine mercy with two days of public prayers in the hermitage of Our Lady of Espinal, but these were not heard and on the third day they experienced this misfortune and with it hunger, misery, disease and death"; 1788, 431)

But not even the power of community prayer could ward off the storm and the widespread suffering of its aftermath. Abbad does not, however, explain why prayers failed.

On the other hand, Diego de Torres Vargas, a church canon, reported how the power of prayer caused a miracle during a hurricane in 1647:

> Un milagro hallo comprobado en el libro de Nuestra Señora de Guadalupe, que fue cuando en la tormenta rigorosa de San Bartolomé, que hubo en esta ciudad, ha mas de 70 años, se llevó el aire una criatura, que pasaba de una casa a otra, y encomendándose a la Virgen de Guadalupe, de allí a tres días se halló viva debajo de una teja. ("I find a miracle verified in the book of Our Lady of Guadalupe, when during the severe storm of San

Bartolomé, which occurred in this city more than 70 years ago, a child was blown up into the air, passing from one house to another, and entrusting it to the Virgin of Guadalupe, three days later it was found alive under a tile"; Tapia 1854, 558)

Religious belief continued to be at least as powerful as any hurricane. The Catholic church itself ordered public reciting of a specific prayer, *Ad repellendas tempestades*, to ward off storms: "Suplicámoste, Señor... que se aleje la malignidad de las tempestades y de los vientos... Dios omnipotente y eterno, que nos sanas castigando y nos preservas perdonando" ("We beseech you, Lord... that the malignity of storms and winds be driven away... Almighty and eternal God, who heals by punishing and preserves us by forgiving"; Salivia 1950, 388).

Earthquakes were routinely blamed on sinful behavior. We see this in governor Felipe de Beaumont 's letter from 1615 telling of an earthquake and storm:

> permitió Dios, por nuestros pecados, darnos un temblor de tierra que parecía que venían las casas al suelo, y a los 12 de dicho mes destruyó casi toda esta isla con un tan gran pies de viento que pareciera ira de Dios, que no dejó árbol que no arrancase por el pie destroncado, y todos los frutos y cañaverales maltratados, de manera que los más de ellos se han perdido. ("God, for our sins, gave us an earthquake that it seemed as if the houses were falling to the ground, and on the 12th of that month he destroyed almost all this island with such great feat of wind that seemed to be the wrath of God, not leaving a tree that was not pulled by the root, and all the fruits and cane fields battered, so that most of them have been lost"; Medrano Herrero 2008, 21)

Religion, superstition, and traditional practices and beliefs were mixed in Puerto Rico's popular culture because people attempted to both explain what was unknowable and to forestall disaster. Teodoro Vidal compiles a subsection of prayers against "atmospheric disturbances" in *Oraciones, conjuros y ensalmos en la cultura popular puertorriqueña* ("Prayers, Incantations and Incantations in Puerto Rican Popular Culture"; 2010, 225–249). He includes prayers and images and artifacts believed to protect against a variety of climatic disasters such as hurricanes, droughts, storms, waterspouts, and thunderstorms. Protective artifacts were also commonly used. Fernando Ortíz notes talismans such as: "[l]as palmas del Domingo de Ramos usadas contra las tempestades suelen exponerse con sendos nudos de sus lacinias" ("[t]he palms from Palm Sunday used against storms are usually exposed with two knots on their splits"; 2010, 59). These can still be seen in Puerto Rican homes.

Vidal also reports popular beliefs to foretell foul weather, for example, a bumper crop of avocados meant a strong hurricane season, many cobwebs falling was a bad sign, as were clouds with reddish hues (which the Tainos believed). The silence of Nature, when roosters and crickets stopped crowing, also was regarded as foreboding. Many movements in Nature were considered as harbingers of doom such as when birds called "rabi-juncos" flew inland. Vidal says that in the western municipalities it was thought the song of the "sanpedrito" (a type of parrot) foretells a storm. As we saw earlier, modern science reaffirms some connections between animal behavior and disastrous weather.

There were enthralling feminine environmental figures both for good and evil: the devil and his witches bring disastrous weather but some female Catholic saints were called upon for protection. Saint Barbara, whose cult can be traced to the ninth century, was a holy being to whom many prayed in times of storms since she was identified with lightning and was a protector against storms. A popular prayer says: "Santa Bárbara bendita, / que en el cielo estás escrita, / con papel y agua bendita; / líbrame de rayos y centellas, / como libraste a Jonás / del vientre de la ballena" ("Blessed Saint Barbara, / who are written in the sky, / with paper and holy water; / deliver me from thunderbolts and lightning, / as you freed Jonah / from the belly of the whale"; Rivera de Álvarez 1983, 78). The quirky connection to a saint "written in the sky" makes this prayer stand out even as it connects to the Biblical story of Jonah and the whale.

Santa Bárbara, as we have seen, was a protector; Vidal quotes three prayers just for her. The generic Virgin Mary as well as Nuestra Señora de Monserrate in Hormigueros were other protective feminine forces (Vidal 2010, 239). Santa Clara also got a share of prayers and devotions (Vidal 2010, 231).

Of course, more masculine than feminine forces were routinely invoked, such as God (identified as male) and Jesus, along with a plethora of Catholic saints, like the Archangel San Miguel, San Isidro Labrador, San Alejo, and San Lorenzo. Vidal points out the deep connections some of these figures and prayers have to other faiths such as Jewish prayers, the Acadian goddess Izar, and a Celtic prayer from the eighth century (2010, 240).

Eighteenth-century master painter José Campeche himself proves the powerful belief in saintly protection. He painted "San Emigdio de Áscoli" in 1787 when an earthquake struck the island. The painting shows the saint floating with angels and to the right of his feet buildings crumble. That year the hurricane and earthquakes caused the collapse of the Arecibo parish and damaged that of Ponce, and San Juan's defensive system was also devastated.

Protest and Exceptionalism

Another early modern recurring theme concerns real-life complaints of governmental inaction, which became a main trope, which has stayed in vogue to the present day thanks to Puerto Rico's perpetually dependent status. This list of grievances about both the lack of effective disaster preparedness and inadequate responses to their destruction appears in 1679, one year after an unnamed hurricane. In a letter written anonymously, the aggrieved writer inaugurates the continuous Puerto Rican stance of grievances about government-led disaster response in ways that have never gone out of style. Fed up with the colonial regimes of don Gaspar de Arteaga (1670–1674) and don Juan de Robles Lorenzana (1678–1683), using a timeless metaphor the anonymous writer links storms and "fatal" government incompetence as: "una rigurosa tormenta, aún más grave y nociva que los que la isla suele padecer en referidos, que le sobrevienen, temporales de agua y viento" ("a rigorous storm, even more serious and harmful than those that the island usually suffers from, as referred to, tempests of water and wind that overwhelm it"; Rivera de Álvarez 1983, p. 37, n. 12). Plus ça change.

The deficiencies of an overlord's disaster response seen in the seventeenth through nineteenth centuries did not abate with the transfer of the island to the United States. Critic Katiana Bagué stresses the U.S.'s deficient response when hurricane San Ciriaco struck in 1899:

> El gobernador militar de Puerto Rico emitió una remisión de impuestos a la isla, pero Estados Unidos no estableció fondos oficiales para el alivio de huracanes. Como resultado, se manifestó un programa de caridad en los Estados Unidos . . . Sin embargo, los fondos de caridad se les dieron a los propietarios de las plantaciones, en lugar de a las víctimas y los que quedaron en la indigencia por la tormenta. ("The military governor of Puerto Rico issued a tax remission for the island, but the United States

> did not establish official funds for hurricane relief. As a result, a charity program was formed in the United States . . . However, charitable funds were given to plantation owners, rather than to victims and those left destitute by the storm"; 2020)

The complaint of misspent hurricane relief remains true to this day. It was raised once more after hurricane Maria against the Congressionally-appointed Financial Oversight and Management Board for Puerto Rico (known as "Junta de Control y Supervisión Fiscal"). Modern-day protests focused on disaster mismanagement, especially the Federal Emergency Management Authority (FEMA)'s many inefficiencies, and the subsequent governmental lack of oversight of the power company LUMA.

Like religion, nationalism is ever a determinant factor in representations of the environment. Patriotic, ecophilic, and exceptionalist depictions of Puerto Rico's environment run counter to the historical and scientific records of disasters. Already in early colonization chroniclers exalted the beauty of the Caribbean bioregion, its lush primordial forests, the beauties of its land, so different from the landscape they knew. Soon afterward, we see forms of patriotism connected to the landscape when powerful feelings often lead to ecoerasure.

Diego de Torres Vargas's 1647 letter already showcases the perpetual garden allegory. Born on the island, his description has an exceptionalist ecoerasing quality. Galen Brokaw says exceptionalism "is most often associated with the political posturing of certain nation-states, but it refers more generally to any discourse that exaggerates the significance of peculiarities in order to set one group apart from, and above, others" (2022, 163). Nineteenth-century Puerto Rican literature will blatantly express exceptionalist fictions of a placid environment. Against all lived and recorded disasters of the time, Torres Vargas describes Puerto Rico's nature as exceptionally mild:

> En el temperamento y calidades [Puerto Rico] se adelanta mucho a todas las islas de Barlovento, porque goza de una perpetua primavera sin que el calor ni el frío llegue a sentirse de manera que aflija ni descomponga la naturaleza . . . Es toda ella fertilísima y verde a la vista de afuera. ("In temperament and qualities [Puerto Rico] is far ahead of all the windward islands, because it enjoys a perpetual spring without heat or cold being felt in such a way as to afflict or alter nature . . . It is very fertile and green to the outsider's gaze"; Tapia 1854, 448)

His exuberant praise of Puerto Rico's environment will become a standard trope of patriotic love of the *terruño* yet, written in 1647, relatively soon after hurricanes in 1641 and 1642, he does not mention storms. He is replying to Spanish Fray Damián López de Haro 's scathing sonnet (of the same year) that called the island backward, partly in response to the devastations of the nameless hurricane of 1642 (Rivera de Álvarez 1983, 34 n. 8). In his response, Torres Vargas gives an exceptionalist, hyped-up description of an Edenic garden. The eminent critic Josefina Rivera de Álvarez says he projects regional pride and a particular admiration for a mythical environment: "con Torres Vargas comienza a tomar cuerpo la manifestación literaria del elogio de la Isla: 'el clima es una perpetua primavera,' 'la vista es variadísima de valles y arboledas,' 'las frutas, dulces y sabrosas'" ("with Torres Vargas there begins to take shape literary praise of the Island: 'the climate is a perpetual springtime,' 'the view of valleys and tree groves is manifold,' 'the fruits, sweet and tasty'"; 1983, 36).

Exceptionalist views of women tie into patriotic mythification and so Torres Vargas rhapsodizes about an important "feature" of the environment: the island's women, already regarded in patriarchal ideology as components of the landscape. Claiming exceptional women as a national commodity was woven into Puerto Rican patriotism; to this day the male gaze views submissive women and a mild climate as the superlative sum of the nation. This ideology is undergirded by unmitigated ecoerasure that ignores the complexities of Puerto Rico's bioregion and regards women as being as decorative as the landscape. Torres Vargas probably was responding to López de Haro 's accusation that women in the colony were so shut-in they didn't even go to mass (Rivera de Álvarez 1983, 34, n. 9). Torres Vargas describes local women as: "las más hermosas de todas las Indias, honestas, virtuosas y muy trabajadoras y de tan lindo juicio que... Todos los hombres prudentes se habían de venir a casar a Puerto Rico" ("the most beautiful in all the Indies, honest, virtuous and very hardworking and with such lovely common sense that... All prudent men should come and marry in Puerto Rico"; Tapia 1854, 478).

Juan Rodríguez Calderón , a Spanish exile, wrote a panegyrical poem, "Hermosa y feliz isla de San Juan de Puerto Rico" ("Beautiful and Happy Island of San Juan, Puerto Rico"), in 1816, despite hurricanes in 1813 and 1814, and two hurricanes in 1816. Regardless of the proximity of disaster, even though slavery was flourishing he overpraises everything about Puerto Rico:

> ¡tus primores! / tu suelo afortunado, / tus nobles moradores . . . tus muchas maravillas . . . el café más selecto, / el azúcar, tabaco / de tinte más perfecto / llenas siempre las eras / de arroz el más sabroso, / de maíz hermoso. ("your beauties / your fortunate soil, / your noble inhabitants . . . your many wonders . . . the most select coffee, / sugar, tobacco / of the most perfect color / the threshing floors always full / of the tastiest rice, / the fullest corn" Rivera de Álvarez and Alvarez Nazario 1982, 55)

He too summarizes the island climate as "eterna primavera" ("eternal springtime" Rivera de Álvarez and Alvarez Nazario 1982, 56) and its people are "apacible gente" ("peaceful folk" Rivera de Álvarez and Alvarez Nazario 1982, 57).

A novel from 1690 features Alonso Ramírez , a Puerto Rican-born protagonist, historiographer, and hero of his own story[9] in which he views the island's environment as "aquella tierra sin escasez" ("that land without scarcity"; Sigüenza 1690, 29). A contradictory text, he adds that hurricanes explain why the "rico" in the island's name was in accurate:

> cuando es cierto que la riqueza que le dio nombre por los veneros de oro que en ella se hallan, hoy por falta de sus originarios habitadores que los trabajen y por la vehemencia con que los huracanes procelosos rozaron los árboles de cacao que á falta de oro provicionaban de lo necesario á los que lo traficaban, por consiguiente al resto de los isleños se transformó en pobreza. ("while it is true the riches that gave it its name were the gold veins that are found there, today because of the lack of its original inhabitants to work them and because of the vehemence with which tempestuous hurricanes razed the cocoa trees that for lack of gold provided what was necessary to those who worked it, consequently the rest of the islanders were thrown into poverty"; Sigüenza 1690, 29)

For Alonso Ramírez , though, the hurricanes are not as bad as the mercantilist economy centering on a supply of abundant gold and on the exploitation to the point of extinction of the natives. The conditions of which he speaks are the intertwined disasters of the emerging Capitalocene: hurricanes plus slavery under a colonial extractive economic system.

Bountiful Land, Endless Sea

The chroniclers had something that we who live in the full consequences of the Capitalocene no longer have, their ecophilic belief in a bountiful Nature that continually self-renews. The process, known as natural succession, posits that ecosystems always regain their balance after disasters. Abbad uses this important part of the ecological thinking of his time, noting several times that hurricanes bring positive processes of renewal for the land in a never-ending cycle of rebirth which nature's violence boosts. In what is an early ecological view, he posits that despite the chaos hurricanes unleash, they are necessary to allow Nature to renew and continue its workings:

> a esta borrasca sucede la serenidad, y á la pérdida de los frutos se siguen las cosas más abundantes, bien sea porque estas violentas agitaciones revuelven los senos de la tierra y preparan su fecundidad, o bien porque el huracán proporciona algunas materias propias á la vegetación de las plantas. Se ha observado que de este desorden resulta una larga serenidad y que la destrucción de los vegetales sirve para su regeneración. ("serenity follows this storm, and the loss of fruit is followed by the most abundant things, either because these violent agitations stir the center of the earth and prepare its fecundity, or because the hurricane brings some of its own materials to the earth. It has been observed that this disorder results in a long serenity and that the destruction of the vegetation serves for its regeneration"; 1788, 430)

The chroniclers most often regarded Nature through the lens of productivity of the land, its bounty as a resource to be used for the benefit of the Crown as well as local well-being. Abbad reaffirms this view of productive renewal when he writes about the storm of 1772 (which was within a cycle of seven storms from 1766 to 1772): "la naturaleza volvió a tomar su curso y la tierra auxiliada del cultivo renovó sus frutos y plantaciones destruidas" ("nature took its course again and the land, aided by farming, replaced the fruit and the destroyed plantings"; 1788, 431). His position has had great staying power. In 1907 writer Félix Matos Bernier compares politics and hurricanes, saying both will pass and life will go on:

> Los fracasos políticos son incidentes pasajeros de la vida de los pueblos. Tienen la importancia eventual de los temporales. Como las tormentas, destruyen, pero sanean el ambiente. Sacuden el organismo social, lo

> perturban un instante, lo golpean, lo hacen vibrar, lo conmueven y, a veces, lo modifican y vigorizan, y pasan. ("Political failures are passing incidents in the life of peoples. Eventually they have the importance of tempests. Like storms, they destroy, but they sanitize the environment. They shake the social organism, disturb it for an instant, strike it, make it vibrate, shake it, and sometimes modify and invigorate it, and move on"; 1907, 11)

Nowadays the notion of a self-renewing environment is highly questionable because of our awareness of the pounding global climate change.

In a different take to the belief in a self-renewing nature, however, a letter in 1642 refers to the land as having been left "sterilized" by storms: "Y de tal suerte [la tormenta] esterilizó llevándose todos los frutales, que el año siguiente de [16]43 padecieron grande hambre" ("and in such a way [the storm] sterilized, taking all the fruit trees, that the following year of [16]43 they suffered great famine"; quoted by Medrano Herrero 2008, 28).

Alejandro Tapia y Rivera: Inflection Point for Environmental Consciousness

Alejandro Tapia y Rivera was a towering intellectual and an anti-colonial writer frequently exiled for his political beliefs. His deeply historiographical *Biblioteca histórica* (complied with collaborators, edited by him and published in 1854) collects many fundamental historical sources about disasters that we cite in this book. These reveal the push–pull of ecophobic and ecophilic views in Puerto Rican culture and also transition into what we call ecoerasure.

Environmental viewpoints in Tapia's *Biblioteca* are important to appreciate Puerto Rico's developing national identity. For instance, he includes a large section of Gonzalo Fernández de Oviedo y Valdés's *Historia natural y general de las Indias* ("Natural and General History of the Indies") which highlights the dire consequences of two hurricanes, San Roque in 1508 and San Francisco in 1526 (1854, 120). Early chronicles present shocking hurricane tales because they were so new to the Europeans and drew the attention of the Crown and aid from its coffers. Many of the historical writings that Tapia collects, however, treat the environment "simply as a silent and passive backdrop to their historical narratives" (Chakrabarty 2009, 204) because they regard humans as paramount. Tapia's patriotism plus a human-centered attitude in many of the texts he collects lead to the book's preamble presenting Puerto Rico as a mild Edenic garden, "la bella soledad de los campos amenos y perfumados de nuestra encantadora Isla" ("the beautiful solitude of the pleasant and perfumed fields of our enchanting Isle"; 1854, 122). This despite the well-documented history of hurricanes and earthquakes—some of which he lived through and which appear in many of the documents he gathers.

Tapia's enormous contributions to images of the Nation often use the theme of eternal springtime, which intensified its hold of the national imaginary in the nineteenth century:

> Atravesada de oriente a poniente por un sistema de montañas, posee valles deliciosos, risueñas colinas y abundantes ríos; y su clima templado y su hermoso cielo y sus campiñas verdes, hacen que Puerto Rico merezca una página en la cartera del viagero y un recuerdo en el corazón del poeta. ("Crossed from east to west by a system of mountains, it has delightful valleys, beaming hills and abundant rivers; and its mild climate and its beautiful sky and its green countryside, make Puerto Rico worthy of a page in the traveler's notebook and remembrance in the poet's heart"; 1854, 3)

Tapia also uses these placid images in his fiction. His story "La antigua Sirena" has an effusive description of an island of "eternal verdura" ("eternal greenery" 1968 206). Tapia's Romantic gaze of *la patria,* his mother(land), is tied to feminine virtues and is explicit in his memoirs when he says, "el amor a la localidad en que se ha nacido es como el amor a la madre" ("love for the place in which one was born is like love for the mother"; 1928, 76). The guiding spirit of this cosmopolitan's recollections (1928, 76) is mostly his ecophilic and ecoerasing regard for the landscape of his beloved *terruño,* and the main synecdoche is the aptly named family home in Guaynabo called *Monte Edén* ("Mount Eden"). He nostalgically labels life there as a "paraíso terrenal" ("earthly paradise"; 1928, 79). Tapia also views the landscape with nostalgia when he laments the long-gone jasmines and cherry trees of his youth: "parece que todo me dice con rumor triste, ya no eres el mismo. Los huracanes nos respetaron, pero los de la vida devastaron tu corazón" ("it seems everything tells me with sad murmur, you are no longer the same. The hurricanes respected us, but those of life devastated your heart"; Aponte Alsina 2015, 17). Saying respectful hurricanes spared his native home gives hurricanes agency but ecoerases disaster.

Francisco Manrique Cabrera says that Tapia's *Mis memorias* is the "spiritual" biography of Puerto Rico (2010, 117), and in that regard Tapia's reverence toward a placid environment is foundational. On the other hand, Tapia, an acerbic critic of the government, constantly focused on government flaws, in contradistinction to the beauty of the placid land. He denounces "tres siglos de letargo y rutinaria ignorancia" ("three centuries of lethargy and routine ignorance"; 1928, 63), in "el funesto sistema colonial absolutista" ("the wretched absolutist colonial system"; 1928, 77). That makes for a greater contrast with the natural environment, which he often rosily describes as benevolent.

Tapia's portrait of a mild climate is incongruous with the reality of Puerto Rican cyclonic ruptures during his lifetime. He was born in 1826 and therefore saw or, when he was in Spain, must have heard of the disasters: hurricane San Jacinto in 1827, a hurricane in 1829, hurricane San Hipólito *and* an epidemic in 1835, the oddly named Nuestra Señora de los Ángeles hurricane in 1837, hurricane San Leoncio IV in 1846, an unnamed hurricane in 1848, then 1850 saw another unnamed hurricane as did 1851, which also saw earthquakes. Later, 1852 topped them with *both* a yellow fever epidemic and hurricane San Lorenzo. How does a patriot reconcile Puerto Rico's violent bioregion with love of Land? Through selective ecoerasure. In 1854 Tapia rhapsodizes about a mild climate while waxing lyrical about beautiful skies in a perpetually green countryside. Not during hurricane season, Alejandro!

Hurricanes are often part of the Puerto Rican cultural subconscious. The real-life climatological context of disaster that Tapia suppresses in the preamble irrupts into his memoirs (written near the end of his life) with descriptions of two hurricanes. He recalls San Narciso's floods and earthquakes, although with the strategic purpose of political commentary on governmental inefficiency, in order to criticize governor Marchessi (cf. 1928, 82). He devotes almost a page to "el huracán de Los Ángeles" ("the hurricane Los Angeles") and its human drama involving a capsized ship and drownings. At the same time, he scores a political point by singling out the infrastructure's lack of potable water in the hurricane's aftermath (1928, 94). Nevertheless, hurricanes rate comparably minor mentions in his memoirs, for instance, he devotes many more pages to *costumbrista* descriptions of dances like the *merengue* and *contradanza*—although here he tactically criticizes the government's silly prohibition of the merengue (1928, 102–104).

Tapia therefore represents an inflection point between Romantic idealizations of the landscape and political denunciations of disaster response within colonialism. When hurricanes appear in the story of the Puerto Rican nation in the nineteenth century it is to underscore human suffering. These disasters feature the human, not the non-human environs: crops and constructions, buildings, and especially churches. The stories are tinged by what Lawrence Buell calls "the ethnocentricity of the human state, where the nonhuman has no voice" (1995, 20).

When the Europeans arrived, they lived a life that was especially hazardous, as we saw in early chronicles. Travel itself, getting to the Caribbean, was perilous in ways that we can now barely conceive. As Bob Sheets and Jack Williams put it:

> Until the middle of the 19th century, anyone who boarded a ship for a journey out of sight of land, was facing a risk of death that no traveler today would even think about accepting. Poorly built ships, inexact navigation, incompetent captains, and inept crews accounted for many accidents, but storms, most especially hurricanes, were the biggest danger. (2001, 21)

The lack of technological advances that Sheets and Williams point to meant that maritime journeys faced untold risks, death of course, but certainly also illness with no way to call for help. This is the context for Eugenio María de Hostos's late Romantic novel *La peregrinación de Bayoán* ("Bayoán's Pilgrimage"; 1863), who imagines sea-bound travel during the protagonist's *peregrinación* from the Caribbean to Spain, and the climate anxiety of the looming threat of storms. During his voyage, high-sea tempests heighten the conditions of Romantic visions of Man (the protagonist) inextricably linked to the forces of Nature:

> el barco se parece a mí: quiere andar, y se lo impiden los vientos contrarios: si acostumbrado a ellos, no pide ya los vientos favorables, empuja velozmente el vendaval: los vientos contrarios lo detienen y lo exponen a la acometida repentina de la tempestad: el vendaval, favoreciéndolo, lo arriesga a naufragar ("The ship is like me: it wants to move, and contrary winds prevent it: if accustomed to them, he no longer asks for favorable winds, the gale pushes rapidly: contrary winds detain him and expose him to the sudden onslaught of the storm: the gale, favoring him, risks shipwrecking him" 1863, 242–243)

Sea storms heighten emotions of fear and dread for Bayoán's voyage of identity as the human life mirrors a stormy sea of tribulations. Tempests bring Romantic drama.

Historian Salvador Brau also centers the human element in his *Lo que dice la historia* ("What History Says," 1893) which follows the style of early modern stories of Puerto Rico and delivers stories of disasters, which is apt since there were hurricanes in 1888, 1889, and 1891. In a traditional manner, however, he pays more attention to local bravery against French piracy than to the reality of devastating hurricanes:

> La población de Puerto Rico, diezmada por la viruela y el paludismo y azotada por ciclones devastadores, se ofrece como cebo fácil á las represalias de los vencidos en Nápoles y el Piamonte. Buques franceses asaltan en 1528, 1538 y 1554 las playas meridionales de la isla, y unos tras otros han de darse á la fuga, ahuyentados por el heroico brazo de aquellos Robinsones anémicos, encariñados con el terruño. ("The population of Puerto Rico, decimated by smallpox and malaria and hit by devastating cyclones, offers easy bait to the reprisals of those vanquished in Naples and Piedmont. French ships raid the southern beaches of the island in 1528, 1538 and 1554, and one after another they are forced to flee, chased away by the heroic arms of those anemic Robinsons, fond of their homeland" 1893)

Brau and other historians often describe a nation subject to uncontrollable environmental disasters but within that include also pirate stories, since man-made threats are more controllable than hurricanes or earthquakes. Tapia also devotes a novel to a fanciful retelling of an iconic pirate's life, Roberto Cofresí. As Francisco Manrique Cabrera says, *El pirata Cofresí* (1876) is far from being a historical novel, adding that the pirate's real life was different almost "in toto" from Tapia's tale (2010, 111). The real Cabo Rojo-born Cofresí has remained ubiquitous in Puerto Rican culture since the nineteenth century; essayist Ramón López calls him "un hombre-mito boricua" ("a Puerto Rican man-myth"; 2000, 105).

Minor writer Ricardo Toro Soler's novella *Huracán. Leyenda puertorriqueña* (Hurricane. A Puerto Rican Legend") which won a literary prize in Ponce in 1886, is a maudlin and generic tale of a Roberto from Cabo Rojo who takes to piracy against the United States after one of its vessels blithely tramples his boat and brings death and ruin to his family. The deep anti-Americanism in the story prefigures the hostility toward that nation after it became Puerto Rico's colonial overlord in 1898. Like so much of nineteenth-century literature, "hurricane" is a metaphor for human emotions or "las tempestades del alma" ("tempests of the soul"; 1897, 34). *Huracán* is also this Roberto's pirate moniker, reaffirming his exaggerated fury and strength.

Can a transnational pirate represent the Puerto Rican nation? Cofresí's hometown of Cabo Rojo would answer, "Yes!" Critic Yolanda Martínez-San Miguel, however, questions the accuracy of such symbolic national attachments explaining that the concept of pirates circumvents:

> the modern and romantic reappropriations of the figure of the pirate as an icon of the free-spirited men and thinkers that configure a predominantly nationalist imaginary in order to explore a more nuanced argument: the pirates are those who questioned the imperial order, but not necessarily to propose a national project, or a communal identity, based on the idea of a singular form of political or cultural belonging. (2014, 20)

So that Cofresí, though combating Spain, was not affirming the Puerto Rican nation but was acting entrepreneurially. Yet the Androcene myth of the free-roaming man who answered to no authority and was nevertheless a patriotic Puerto Rican persisted well into the twentieth century. In 1938 poet Evaristo Ribera Chevremont waxes poetic about heroes and pirates in "Mar Caribe" ("Caribbean Sea") identifying the sea itself as: "Mar levantado en héroes, mar ronco de piratas" ("Sea elevated by heroes, hoarse sea of pirates"; Quintana 1996, 289). He affirms control over Nature in a patriarchal society that views even the sea as subsumed to heroic men. Chevremont forgets that hurricanes are much louder than humans.

In sum, from the 1500s to the early 1900s prevalent stories of Man against the environment offer solace against the overwhelming power of hurricanes that were barely understood. This starts to change in the middle of the nineteenth century as hurricane-tracking technologies allow more precise measurements of trajectory, strength, and multiple impact points of hurricanes so that we "see" them more clearly. But the question remains, does that lessen our fear of them?

Human-centered narratives from the early colonial period were still fundamental to historical writing even in the twentieth century; for instance, Fernando Picó's *History of Puerto Rico* (updated 2017), signals the primacy of the human element in its subtitle, "A Panorama of its People," and only mentions hurricanes during the early colonial times to explain the significance of the indigenous god Juracán. In the first part of this history, he doesn't, for instance, even allude to the three storms in 1530. Picó concentrates his prodigious scholarly knowledge about the workings of the Spanish colony on geopolitics and the economic impact of the colonial administration—things like the *situado* subsidy and the construction of infrastructure fortifications in San Juan. He neglects to mention the impact of hurricanes on the economy and only glancingly refers to how the construction and foundation of towns factored in floods.

As to human threats, Picó's *History* has highly entertaining stories of sixteenth- and seventeenth-century attacks on San Juan by English and Dutch corsairs—including a catastrophe in 1625. Picó mentions one non-human disaster, a storm during the early Spanish colonization when the shipping system of *flotas*, gave corsairs the ability to attack "naves dispersas por huracanes" ("ships scattered by hurricanes"; 2017, 78). As noted, pirates make for manly adventure tales, so Picó includes stories of the pirates Drake and Cumberland. Denouncing bad colonial governance that made disasters worse, he explores other calamities, for instance, the epidemics of the seventeenth century brought on by colonization in the Caribbean. He includes the cholera epidemic of 1855 that left 30,000 dead. Brau explores how poor infrastructure allowed the growth of shantytowns which resulted in the spread of the epidemic. His history can include disasters if they advance a story of the Nation overcoming adversity. We will see in the coming chapter how ecoerasure reimagines a prettified environment, ignoring the "ugliness" of real ecology and meteorology in order to define and strengthen the story of the Nation.

Notes

1. Fan and Liu identify other technologies: "single geologic proxy to multi-proxy techniques by integrating microfossils, sedimentary organic elemental ratios, and stable isotopes, together with typical sedimentary textures and structures, for the diagnosis of storm deposits" (2008, 2907).
2. Scientific studies such as these enhance our historical understanding of climate, however, as Braun et al. note, the knowledge gained by paleotempestology often addresses the infrastructure needs of the present: "the history of tropical cyclones is important for conducting a risk assessment for coastal communities that are susceptible to hurricane strikes" (2017, 360).
3. Figures shaped like the letter S.

4. Librarian of the Free Public Library of San Juan who published the first major historical study of Puerto Rico in English in 1903.
5. For a summary see Nicole Caldwell.
6. José Luis González says it is the first historical account by a native-born resident (1976, 30).
7. There were three.
8. He studies Cuba, but most likely similar for the Tainos of Puerto Rico.
9. The book was penned by Carlos de Sigüenza y Góngora and there is still discussion if it is fiction, historical document, or testimony. In the 2018 edition, editor José F. Buscaglia-Salgado offers proof that Ramírez was a historical character.

CHAPTER 3

Nation-Building Ecoerasures of Disaster

The nineteenth century opened with hurricanes Santa Rosalía and San Mateo (both in 1804) and closed with San Ciriaco in 1899, a punishing Category 4 monster. In addition, there was the metaphorical storm of the U.S. takeover in 1898. 1899 had two earthquakes. Why, then, are "natural" disasters underrepresented in literature and art in nineteenth-century and early twentieth-century depictions of the environment? The disaster data between 1804 and 1899 in our Appendix X shows a terrifying environment. We will see how stories, images, and symbols of the landscape link to patriotic nation-building, as nostalgic and bucolic depictions enhance love of land and bolster nationalistic pride. Ecoerasure is our term for the process by which culture "forgets" the disasters of its bioregion and portrays only the beauties of non-lethal moments in the Caribbean environment. All of which gave us canonical works of literature and art that domesticated the landscape, rendering it fit for patriotic and commercial consumption. Yet, as we will see, like the suppressed pain and trauma of a collective national unconscious, stories, images, and symbols of disaster blast through into a few cultural products—exceptions that prove the rule that the environment in Hurricane Alley can never be wholly ecoerased.

Furthermore, the burden of representation falls on women as they are depicted as placid counterparts to a fictionally serene environment.

M. Acosta Cruz, *An Ecocritical Study of Puerto Rican Culture*,
https://doi.org/10.1007/978-3-032-04044-2_3

Images of mild women and a placid environment are therefore significant props to the nation's colonial and patriarchal socio-political realities. These fictitiously serene configurations are replicated everywhere in art and literature in the nineteenth century.

The unfulfilled dream of Puerto Rican independence enhanced emotional attachments to a land imagined as placid. Given Puerto Rico's political dependency it is logical that the landscape and climate were portrayed in the most patriotic, favorable way during the time of most intense national identity formation. Ernst Renan famously said that "[f]orgetting, I would even say historical error, is an essential factor in the creation of a nation," so that in Puerto Rico's case forgetting disaster help build a shared national identity. These ecoerasures were not, of course, universal, and we see some competing interpretations of Puerto Rico's environment, sometimes within the same works, as Tapia's vast *oeuvre* shows. Nonetheless, most works of history, literature, and art were usually well and truly tied to love of a bucolic landscape, in part because, "the enlistment of the pastoral in the service of local, regional, and national particularism [is] an article of cultural nationalism" (Buell 1995, 32). Also, a great many works depict a rural *patria* and turn away from cityscapes, which often means ignoring the expansion of landscape-destroying infrastructure in cities like San Juan.

Exceptionalist literary and artistic representations of a harmonious, eternally unchanging Nature are partly brought on by the belief in the ecological process of natural succession because Caribbean foliage regrows quickly after hurricanes. As discussed in Chapter 1, the exceptionalist view of a serene homeland environment did not begin in the nineteenth century; for example, already in 1765 field marshal Alexandro O'Reylly praises the produce the island contributes to the Crown, everything from fish and water, to coffee, ginger, corn, and he hyperbolically applauds its sugarcane as the tallest and sweetest in America (Tapia 1854, 516). He lauds "esta preciosa isla" ("this precious island"; Tapia 1854, 524) and mentions hurricanes only as a time-keeping measure (along with visits from the bishop), praising Puerto Rico for its large number of Caucasian (if lazy) people. That focus on the productivity of the land will become the norm. As O'Reylly proves, the myth of a serene landscape did not appear suddenly, other depictions of a bucolic paradise at the beginning of Puerto Rico's culture, however, are most often intermixed with realistic appraisals of the harsh climate conditions the chroniclers faced. For example, Juan Ponce de León II (Troche) and Antonio de Santa

Clara's *Memoria de Melgarejo* ("Memoir of *Melgarejo*" from 1582 in the same sentence says that the temperament of the island is very good and then notes "except" for hurricane season (110–112).

The aestheticization of the landscape benefitted from significant gaps in cyclonic and geological activity between disasters in both the nineteenth and the early part of the twentieth centuries; for example, the gap from 1855 (cholera) to 1867 (a terrible year which brought both earthquakes and hurricane San Narciso). Similarly, the years between 1876 and 1888 have no island-wide disasters. 1899 to 1916 marks another hiatus. Later, the island got a long break from significantly destructive hurricanes from 1932's San Ciprián to Santa Clara/Betsy in 1956. The latter was a tremendously fruitful period in Puerto Rican culture.

The myth of a placid landscape in Hurricane Alley corresponds to what Alison Byerly calls "the wilderness myth," the impulse to domesticate the landscape, along with an aversion to dwelling on traumatic disasters. Byerly discusses the loss of this wilderness myth in relation to the U.S. landscape, saying the sublime vista that "staggered" early settlers was "gradually reduced and circumscribed until it no longer seems to stretch into infinity, but is contained and controlled within established boundaries" (1997, 53). The same cultural domestication of Nature happens in Puerto Rico. Patriotic love of country brings ecoerasure to the *terruño*, depicted as a dainty land where everything is placid. The environment is also portrayed as a passive background in which humans dwell, where the non-human has little to no agentive presence.

Emphasis on the human, i.e., dismissing or overlooking the non-human, is a key component of nineteenth-century ecoerasure. A fictional environment as a "slow and apparently timeless backdrop for human actions" permeates the culture of the time (Chakrabarty 2009, 205). This mindset is also linked, as William Flores says, to Christian views of the natural world as unlimited and permanently renewable, the "Argumento Dios proveerá" ("Argument that God will provide"; 2015, 63).

No culture is a monolith, so even in the nineteenth century there are works in which hurricanes are part of the plot; for instance, Matías González García, a minor writer, uses a hurricane as one of many negative elements in a human-centered tale of a *jíbaro*'s hardscrabble life.[1] The character's woes worsen after a hurricane: "un terrible huracán acabó con la cosecha, y el pobre labrador se fue afligidísimo al pueblo, a decirle a su compadre que aquel año no le podría pagar" ("a terrible hurricane wiped out the harvest, and the poor farmer went to the town in great distress, to tell his *compadre* that he could not pay him that year"; Aponte Alsina 1984, 127). Again, Nature is valued for the harvest, for its socio-economic impact; the story is not interested in the hurricane destruction of the non-human environment such as landslides, decimation of flora and fauna, and redrawn coastal lines and riversides.

An ecoerasing literary tradition truly blooms in the nineteenth century; María Babiana Benítez, Puerto Rico's first woman poet, shows an earnest neoclassical certitude in regarding the nation's environment in Edenic terms—which we will see repeated throughout the nineteenth and twentieth centuries. Benítez's 1832 poem "La ninfa de Puerto Rico a la Justicia" ("The Nymph of Puerto Rico to Justice") predates the *Aguinaldo puertorriqueño* ("Puerto Rican Christmas Song/Gift"), the 1844 volume regarded as the first Puerto Rican literary anthology. Benítez extols a pastoral vision of the proto-nation: "Yo, desde Luquillo en la alta cima, / contemplo mis riberas, / que Ceres misma anima, / cubiertas de frondosas sementeras" ("I, from Luquillo's high summit / contemplate my riverbanks, / which Ceres herself animates, / covered with lush fields"; Rivera de Álvarez and Álvarez Nazario 1982, 67). We see here a solidifying bucolic environmental imagination centered on a myth of a productive land in a fictionally peaceful climate. Equally important, ecocritically, is that Benítez only pictures the non-human insofar as it benefits mankind; in the next stanza she domesticates all the "fieras" ("beasts") that are "amigas del hombre" ("friends of man").

Hurricanes, however, were a notable enough occurrence in Benítez's time. She writes her eclogue five years after hurricane San Jacinto, which, granted, went through the southern and western parts of the island and her town of Luquillo is in the northeast, so maybe she did not see that hurricane's effects. Earlier though, in 1825, hurricane Santa Ana had caused 374 deaths, and is estimated to have been a monstrous Category 4. Benítez and her later peers, nevertheless, choose ecoerasure time and again to depict land and sea as welcoming so Puerto Rico—and by extension the Caribbean—becomes a pastoral setting. Their choices will become entrenched in Puerto Rico's cultural traditions.

Benítez's environmental imagination begins a pastoral nationalism that develops fully in the nineteenth century and persists until roughly 2017. Throughout the poem she extols this pastoral vision of eternal spring. Significantly, she specifically links the land to feminine energies:

> en el ameno campo do reflecta / de terna primavera claro día . . . / Aquí crece la espiga vanidosa / de llevar bien temprano / la mies, dulce, abundosa / que lo mismo en invierno que en verano, / recoge la zagala primorosa ("In the pleasant landscape which reflects / the clear day of an eternal spring . . . / Here grows the vain sprig / that takes early on / the harvest, sweet, abundant / the same in winter as in summer, / that the exquisite young woman picks"; Rivera de Álvarez and Álvarez Nazario 1982, 66).

A placid and plentiful land interpreted as feminine or through feminine energies will become a recurring topic in the nineteenth century, to the detriment of realistic portrayals of women and the environment. Her vision also includes a related trope, the nation as an infantilized Land.

Josefina Rivera de Álvarez notes the literary and political implications of Benitez's poem in connection to the rest of the nineteenth century: "las impresiones que nos comunica sobre la naturaleza y el paisaje de Puerto Rico encuadran en el marco... del XIX: campiñas y horizontes de ideal y apacible belleza, ambiente idílico... de grandilocuencia expresiva" ("the impressions it communicates about the nature and landscape of Puerto Rico fit into the framework... of the nineteenth century: countryside and horizons of ideal and peaceful beauty, idyllic atmosphere... of expressive grandiloquence"; 1983, 109). The move to imagine a fictional environment entirely free from disaster is deeply intertwined with a frustrated patriarchal and colonial mentality which wants control over both environment and women.

Picturing Nature as beneficial seems to say, yes, Spain (and later the United States) may rule us, but a placid, benign Nature favors us. The Land—often extending to the seas—supports patriotic love for a beautiful, peaceful *Isla*[2] regarded as a fount of only goodness. The sense of national belonging, of ownership of the Land, is reinforced by a shared vision of the land and its people as unmarred by the scars of disaster trauma.

Literature was one of the key nation-building ingredients in the birth of Latin American national identities, Puerto Rico included. Puerto Rican cultural identity is not in question despite its political dependency as the author's book *Dream Nation* proves. Importantly, its national, and nationalist, literature as such develops during the 1837 to 1846 gap in hurricanes, with the publication of the following five foundational texts.

What kinds of depictions of the bioregion's environments do we see in them? Let us examine them one at a time.

I. Francisco Pastrana, Mateo Cavailhon y Alejandrina Benítez, et al. *Aguinaldo puertorriqueño* ("Puerto Rican Christmas Song/Gift"), 1843 (anthology of poems and short stories). Published during the hurricane gap from 1837 to 1846.

Aguinaldo puertorriqueño has a precedent-setting heap of ecoerasure. Yet the authors write six years after 1837's hurricane Nuestra Señora de los Ángeles "caused exceedingly grave losses of life and property. All the ships in the harbor of San Juan were lost" (Middledyk 2022, 220). Another important circumstance is that 1837 also dashed hopes of autonomy or sovereignty when the Spanish Courts dictated special laws for Puerto Rico and Cuba and, again in 1838 when the Utuado/Ciales uprising was squashed.

Despite the *Puertorriqueño* in the title the writers include mostly descriptions of non-Caribbean environments. Instead, they look to Europe because that is the literary tradition they follow. One does not have to read far into this volume to see the guiding nationalist and, equally important, heteronormative principles of the writers, a mix of Spaniards, a Venezuelan and six Puerto Ricans, whose volume is of historical significance but (critics agree) little literary depth. The *Aguinaldo* writers signal their patriotic zeal saying they write, "animados del mas ardiente celo por todo aquello que pueda contribuir a dar lustre y gloria a esta venturosa Antilla" ("animated by the most ardent zeal for everything that can contribute to give luster and glory to this fortunate Antille"). The postscript equally states their intent to make the island well-known (1843, 199).

Two elements, a placid land and meek women, appear throughout the volume within the backdrop of an imaginary land free of hurricanes and earthquakes. The first quartet of the book's introductory poem has two elements that are linked to the Nation: an idealized life of "sendas de flores sin ninguna espina" ("paths of flowers with no thorns"; 1843, 11) and beautiful women (likened to such things as flowers and little butterflies across the contents). As poet and scholar Juan Pablo Rivera says about the first poem of the anthology (which applies to the entire volume):

> la tradición poética puertorriqueña entera se origina con esta ceremonia de seducción heterosexual que revela las mores y expectativas de género en

> la colonia durante el periodo de forja de una incipiente identidad nacional ("the entire Puerto Rican poetic tradition originates with this ceremony of heterosexual seduction that reveals the mores and gender expectations in the colony during the period of forging an incipient national identity"; 2021, 21).

Heterosexual love and a placid landscape are inextricably linked in the anthology and will remain in place for a very long time. Romantic heterosexual love is a guiding principle for most of the poems; Josefina Rivera de Álvarez calls their style "Romanticismo de imitación" ("imitative Romanticism"; 1983, 133). Nostalgia is also a common tenet of the volume, with the perception that the golden age was always past, that earlier times were ever better. As a result, apart from the introduction and the postscript, the *Aguinaldo* entirely avoids the bioregion's unsettling, if not frightful, climate and geology. On the contrary, the stories and poems are set in foreign lands. The first story, "Pedro Duchateau" by Martín Travieso, is set in Seville. The second piece, a poem about "Oriental" dreams, "with 100 beauties" in crystal palaces, is set in some unspecified land. Another story, "Historia de Don Alfonso de Cordoba" ("Story of Don Alfonso de Cordoba") is also set in Spain. The stories and poems all show strong ties to many principles of European Romanticism, with its representations of medieval times and "exotic" women. For instance, the story "El astrólogo y la judía. Leyenda de la Edad Media" ("The astrologer and the Jewish Woman. Legend of the Middle Ages") by Mario Kohlmann is a travelogue featuring Bagdad and Cairo, but not San Juan nor San Germán.

The writers of *Aguinaldo puertorriqueño* missed out on the Romantic shock and awe that could have resulted from just one depiction of the sublime terror of Caribbean hurricanes. Their distorting ecoerasure of the environment leans instead to picturesqueness, a quaintness that belies climate reality and domesticates the landscape for patriotic consumption. Yet these Puerto Rican writers seem unable to do without storms within their cultural subconscious as fictional storms bring drama and drive some of the works. When storms appear, however, they not set in the Caribbean and are clichéd metaphors for life's misfortunes. Hurricanes feature in the volume as dramatic background attached to no specific location, certainly not Puerto Rico or the Caribbean, neither of which are portrayed. For example, Kohlmann's story features two hurricanes, one surrounds the cave in which the astrologer/mage tries to work his magic. Is this a real hurricane? No, it seems to come from the devil-tinged convulsive power used by the mage, but the experience of hurricanes (torrential rain, howling winds) was that of some of the real authors, even if that story is set in the Middle East. The story features a storm at sea, in which the drama of the natural elements mirrors the characters' suffering, a common Romantic trope.

Storms are there to heighten tension and move plots. The poem "A Armanda" ("To Armanda") features a storm at sea that serves as stock background and as a motif to imperil the leading lady. In the poem "A un laurel" ("To a Laurel") a hurricane's power brings down the symbolic laurel as the storm hits, representing a harsh fate, "la suerte dura" ("harsh luck"; 1843, 129), in the garden of life. Storms often provide atmospherics for Romantic visions, but they are not real descriptions of actual climate phenomena. The poem "A Jacobo" ("To Jacob") uses the trope of life as a ship in which the hurricane is, of course, representative of sudden misfortunes that can strike at any time, as unexpectedly as hurricanes struck back then.[3] In this poem the storm exists to hurt the poet's heart: "de improviso / no has visto negras nubes apilarse / y el feroz huracán tras ronco aviso / al ancho seno con furor lanzarse" ("Suddenly / have you not seen black clouds pile up / and the fierce hurricane after gruff warning / throw itself on the ample chest with fury"; 1843, 191). Interestingly, this poem also uses the belief in natural succession we have previously explored, in this case how the poet's heart, here symbolized by a leafless tree, blooms again (1843, 192). Furthermore, the poem signals this renewal by the shining of a star (the lady love) in the middle of the storm. So, *Aguinaldo Puertorriqueño* is a curious proto-nationalist volume that ecoerases the local landscape. That will soon change.

II. Manuel A. Alonso, Santiago Vidarte et al. *Album puertorriqueño* ("Puerto Rican Album"). 1844 (anthology). Also published within the hurricane gap from 1837 to 1846.

These authors, all students in Spain, are full of earnest sincerity and their volume brims with Romantic themes with the same ecoerasures as in *Aguinaldo Puertorriqueño*, with no regional nature nor disasters. Like the earlier volume, the writers express love for their homeland, "nuestra patria," which leads to some local color, although this volume too relies on European landscapes, as well as on imitative Romanticism.

Women are again components of Nature; a poem by Pablo Sáez, for instance, directs its exceptionalist gaze at the beauty of "La puertorriqueña," ("The Puerto Rican Woman"). The volume's *costumbrismo*[4] highlights local mores in Alonso's "La fiesta de Utuao," ("The fiesta in

Utuao"). Nostalgia is again a strong element, Santiago Vidarte's nostalgic poem "Un recuerdo de mi patria" ("A Memory of my Homeland") presents the island as an inspirational memory.

III. Manuel A. Alonso. *El Gíbaro* ("The Gíbaro"). 1845 (anthology). Also published within the hurricane gap from 1837 to 1846.

Alonso's *costumbrismo* continues the expansion of national identity now deeply rooted in a regional landscape portrayed as mild and always welcoming. Typical of this style is the story "El pájaro malo" ("The Bad Bird") which offers detailed, loving descriptions of a landscape that literally shelters the traveler:

> Si el lector ha hecho alguna vez el camino de Caguas a la capital de Puerto Rico, recordará el hermoso valle que media entre la cuesta de Quebrada-Arenas, y el cerro llamado de la Mesa; valle ameno y muy fértil regado por el río Cañas y la Quebrada-Arenas, y sembrado de infinidad de árboles, algunos de los cuales, situados a la orilla del camino, sirven de día para guarecerse el viajero de los ardores del sol ("If the reader has ever gone on the road from Caguas to the capital of Puerto Rico, he will remember the beautiful valley that runs between the slope of Quebrada-Arenas and the hill called de la Mesa; a pleasant and very fertile valley watered by the Cañas River and the Quebrada-Arenas, and planted with countless trees, some of which, located on the edge of the road, serve during the day to shelter the traveler from the heat of the sun"; 1845, 175)

Despite the sheltering trees, "El pájaro malo" is a story full of violence, and its moralizing intent defines the qualities of the *jíbaro* as morality, honesty, and hard work, like the humble man Perico Paciencia from the homonymous story.

This fundamentally moralizing tone for national identity is resonant with a view of people's goodness linked to a placid, storm-free milieu. Alonso, however, uses a metaphor from technology: "el verdadero termótetro del estado de civilización de un pueblo" ("the true thermometer of the state of civilization of a people" (1849, 94). This kind of technological metaphor will become more prevalent.

IV. A second *Aguinaldo Puertorriqueño* appears in 1846, with almost all the same authors of the previous one and employing the same topics and Romantic idiosyncrasies as the first *Aguinaldo*. Even though it was published the same year hurricane San Leoncio struck, it may have been written before that hurricane on Sept. 12, which may explain why no hurricane mars the anthology's nostalgic and patriotic tone.

V. Manuel A. Alonso, Santiago Vidarte et al., authors of *Album puertorriqueño*, published *Cancionero de Borinquen* ("Songbook of Borinquen") in 1846. Many of its texts feature Romantic elements seen in the earlier anthologies: Orientalism, European settings, romantic love, and similar topics. Here, as in the earlier collections, the writers deploy a traditional trope: the mother as sentimental cultural touchstone.

The *Cancionero*, however, now has depictions of regional nature, overcoming earlier reliance on European settings. Local flora and fauna appear often as symbols tied to the nation's uniqueness; for example, Francisco Vassallo separates local fruit from foreign fruit, mocking a man who prefers European peaches or pears whereas Vasallo favors the homebased "agridulce guanábana" ("bittersweet soursop"; 1846, 51). Local specificity will fully bloom in art with Francisco Oller y Cestero's magnificent still lives of local produce, including *guanábanas*. Pablo Sáez invokes nostalgia for the local milieu, using sugarcane fields, parrots, hummingbirds, and the *pitirre*—a small bird that becomes symbolic of Puerto Rican pluckiness (1846, 199).

The Garden Isle

Nineteenth-century literature recursively uses the tenacious allegory of Puerto Rico as a garden. First, the garden references Biblical symbolism of purity and God's harmonious environment. The 1903 version of Puerto Rico's anthem (replacing Lola Rodríguez de Tió's revolutionary version) centers on the concept of a "jardín florido / de mágico esplendor" ("flowered garden / of magical splendor") elevating the environment to a magical realm.

Michel Foucault says the garden is perhaps the oldest example of a heterotopia: "a sort of simultaneously mythic and real contestation of the space in which we live" (1970, 232). The nineteenth-century Puerto Rican cultural enterprise fully deploys the garden heterotopia with its "symbolic perfection" (Foucault 1970, 234), ecoerasing chaotic hurricanes and earthquakes. Foucault's definition is especially on point for us since "the garden has been a sort of happy, universalizing heterotopia" (1970, 234). Instead of furious winds and relentless rains, the writing of the times depicts the cheerful picturesque garden environment, showing what John Dixon Hunt calls a "determination to process the messy world" (1992, 6). Foucault says that the role of heterotopias is to create "a space that is other, another real space, as perfect, as meticulous, as well arranged as ours is messy, ill constructed, and jumbled" (1970, 235). Reality is jumbled indeed after a hurricane or an earthquake.

Ecoerasure needs the garden allegory as a heterotopia to encompass the will to order, to control Nature, which may have been possible in European gardens but had and has severe real-life limits in the Caribbean. Examples in the *Cancionero*: the first poem calls Puerto Rico "bellísimo pensil" ("most beautiful delicious garden"; 1846, 6); Vasallo in "A Borinquen" ("To Borinquen"), looks at the island from the sea, pointing out that PUERTO RICO (in caps) is a "fantástico jardín de flores rico" ("fantastical garden rich with flowers"; 1846, 29); Santiago Vidarte categorically calls it "nuestro precioso Edén" ("our precious Eden"; 1846, 234).

The apotheosis of the Edenic island as a garden of unsullied beauty will later in the century be José Gautier Benítez's poem "Puerto Rico" (from 1879), a poem suffused with bucolic imagery. His ecoerasing depiction won a prize from the Ateneo Puertorriqueño three years *after* hurricane San Felipe I,[5] which caused much destruction and a total of 19 deaths in 1876. And yet the main allegory about the nation is again the tranquil garden! The first four lines, memorized by generations of Puerto Rican school children, are:

> ¡Borinquen!, nombre al pensamiento grato / como el recuerdo de un amor profundo, / bello jardín de América el ornato, / siendo el jardín América del mundo ("Borinquen!, a name pleasing to one's mind / like the memory of a deep love, / a beautiful garden, the ornament of America, / America being the garden of the world"; Rivera de Álvarez and Álvarez Nazario 1982, 167–169).

Gautier Benítez underscores a recursive cultural trope: the smallness and daintiness of Puerto Rico. Throughout this ode to Puerto Rico the poet highlights the mildness of flora and fauna using a denaturing frame for Nature, specifying that no tigers, lions, jaguars, boas, or caimans mar the *patria*'s environment. Nothing in his portrayal of the Puerto Rican environment is big or extravagant, on the contrary, he says no volcanos nor Niagaras distort the horizon in which the palm, that symbol of the Nation, placidly sways. Wither San Felipe? Even when he looks at the land from the sea it is through a "bruma," a softening sea mist. Gautier Benítez, however, does not deny the wrongs of history, he does not soften or mitigate the damage of one brutal aspect of history, "la vil esclavitud manchó tu suelo" ("vile slavery stained your soil"; Rivera de Álvarez and Álvarez Nazario 1982, 168).

Nineteenth-century writers like Gautier Benítez imagine island and woman as the selfsame, usually in a diminutive key. For instance, his poem "A. C. B." equates garden island and virginal woman: "Mi dulce niña puertorriqueña / luz de mis ojos, linda trigueña / vergel de gracias, cuna de amores... Tu talle es cuál las palmas americanas; / tu acento, cual la brisa de mis sabanas" ("My sweet Puerto Rican girl / light of my eyes, pretty brunette / garden of graces, cradle of love... Your form is as American palm trees; / Your accent, like the breeze from my savannas"; Manrique Cabrera 2010, 122). Much of the literature we are examining here was written under the patriarchal gaze of men in a society that infantilized women and the landscape to replicate the then-standard sexist softening of all feminine entities. There is an explicit voluptuousness and femininity to Gautier Benítez's landscape, with more than a hint of the femininity of virgin jungles. Marta Aponte Alsina specifies the connection in his poetry to "un código de la isla / virgen / jardín" ("A code of the island / virgin / garden"; 1984, 93). It is very much a prettified land, where no disaster mars the landscape.

Gautier Benítez mentions hurricanes twice; first as a (fictionally) forgiving force that spares Puerto Rico, allowing tranquil ships to reach the Caribbean Garden. Later he uses the ship metaphor to state the *patria* will not be the warship that defies a hurricane, his figuration for the island is instead a placid little boat. With his comparison Gautier Benítez proves how common it was (and is) to misrepresent the strength of cyclones. Ships could not, and still cannot defy a major cyclone, as the battleships of the U.S. Third Pacific Fleet discovered to their regret in 1944. Kerry Emanuel illustrates the destructiveness of cyclones against warships citing harrowing accounts of two typhoons that struck the Third Fleet, sinking many warships and killing hundreds of men; Admiral Halsey, commander of the Fleet, writes:

> No one who has not been through a typhoon, can conceive its fury. The 70-foot seas smash you from all sides. The rain and the scud are blinding; they drive you flat-out, until you can't tell the ocean from the air. At broad noon I couldn't see the bow of my ship, 350 feet from the bridge. The *New Jersey* once was hit by a five-inch shell without my feeling the impact; the *Missouri*, her sister, had a kamikaze crash on her main deck, and repaired the only damage with a paintbrush; yet this typhoon tossed our

> enormous ship as if she were a canoe. Our chairs, tables, and all loose gear had to be double-lashed; we ourselves were buffeted from one bulkhead to another; we could not hear our own voices above the uproar. (Emanuel 2005, 180)

In Gautier Benítez the small boat of the *patria*, in its minor exceptionalist key coupled with Romantic cliché, ecoerases the reality of the climate that had unleashed a hurricane three years prior to his writing, but by the time he writes in 1879 the placid island allegory already had become ingrained in Puerto Rican culture.

Returning to the corpus of nineteenth-century literature, the majority of works that use hurricanes do so as metaphors for romantic feelings. Even as late as 1895 Ana Roqué deploys one of these tired metaphors: "los labios temblorosos como si un huracán de pasiones indomables rugiese en el fuero interno de aquel ser" ("lips trembling as if a hurricane of indomitable passions roared in the inner self of that being"; Aponte Alsina 2015, 75).

Nature is, of course, placid throughout the *Cancionero*, "Ya del campo en la plácida verdura / Ya del cielo en el limpio pabellón" ("Or from the countryside in the placid greenery / Or from the sky in the clear pavilion"; 1846, 169). Real-life storms, however, unsettle these Puerto Ricans. Santiago Vidarte, for instance, uses the hurricane for its religious significance—something we have explored in Chapter 2—as he begs God to intervene in the face of a hurricane that approaches because of his sins, a complete throwback to early colonial beliefs (1846, 139).

A poem titled "La tormenta" ("The Storm") by Ramón E. de Carpegna, clearly does not reflect Puerto Rico's actual environment, instead the backdrop is a mishmash of fanciful disasters: in the middle of the howling storm there is also a volcano (1846, 215). The hurricane is again a symbol of life's travails, which in this poem is also an excuse for the poet to soothe the timorous beloved woman "mi ninfa amada" ("my beloved nymph"; 1846, 216). Rather than depicting Puerto Rico's environment, the collection uses an entirely fictitious Nature; Francisco Vassallo makes note of "ese atroz berenjenal / que la gente llama mundo" ("that atrocious mishmash / that people call the world"; 1846, 35) this is when his own *berenjenal* lumps together animals that do not co-exist in the natural world: wolves, sharks, caimans, elephants, and a donkey.

There is a palpable, though kitschy, love for the region in the *Cancionero*, as it lays the beginnings of a cluster of environmental symbols of the authenticity of the *patria*. Many texts in the *Cancionero* use the palm tree, one of the most consistently deployed symbols of national identity. In Pablo Sáez's poem "La palma" ("The Palm Tree") the palm stands tall even against hurricanes: "alza tu frente, magestuosa palma... No temas, no, que el huracán te rompa" ("Raise your brow, majestic palm... Do not fear, no, that the hurricane will break you"; 1846, 11). The hurricane, here a symbol for colonialism, is then displaced and distorted, wrenched from its place in the Caribbean to become an evil similar to the environment of "tierras heladas" ("frozen lands"; 1846, 13) of, presumably, Spain. The palm is, of course, the Nation.

And yet contradictorily, more often these Romantic writers shy away from even mentioning hurricanes and earthquakes as actual lived dramatic phenomena. This development is peculiar given that they were imitating the Romantic imagination of Europe and the United States, which often contrasted humans with the vast scale of a grandiose Nature for emotional effect. Romantic creators frequently expressed feelings about life-altering events through tempest imagery. Eighteenth-century Germany's proto-Romanticism with its *Sturm und drang* would have no meaning without reference to turbulent storms, both inner (the spirit) and outer (Nature). European Romantics swooned over the majesty of storms, for instance, in *Childe Harold* Lord Byron's *cri de coeur* calls out to a feminine storm: "O night, / And storm, and darkness, ye are wondrous strong / Yet lovely in your strength as is the light / Of a dark eye in woman!" (*Canto III* 1894, 1–4). Wordsworth tied his inner turbulence to the outer storm: "I walk alone / in storm and tempest" (*The Prelude* 2004, 2.321–2). Moreover, the southward journey in Coleridge's *The Rime of the Ancient Mariner* begins with a "storm-blast" (1857, 10). *Moby Dick* in 1851 would be very different if there were no storms, or Nature, to battle.

It is therefore striking that early Romantic Puerto Rican writers, who had many more direct experiences of real—not metaphorical—storms, did not avail themselves of the actual environmental distinctiveness of their homeland but contrariwise, pictured their Land as mild. Later in the century Alejandro Tapia y Rivera's late Romantic bravado uses storms, but only so that the poetic voice can overcome through the counterforce of emotions and determination, "¡olas enfurecidas, vientos desenfrenados, nubes tenebrosas, relámpagos de luz, truenos y rayos, en nada os temo" ("angry waves, unbridled winds, gloomy clouds, bright lightning, thunder and lightning, I fear you not"; 1968, 374). The Androcene image of man vs. tempest also appears in his *El pirata Cofresí*, "Él, que no perdía ánimo ante la tempestad, porque ésta tenía algo de su ser" ("He, who did not lose heart in the face of the storm, because it had something of his being"; 2015, 35). Cofresí may not have lost heart, but he did lose a ship during the nameless hurricane of Sept. 9, 1824 (Miner Solá 1995, 22). Yet Tapia still links storms to human emotions.

Exile, another nineteenth-century theme, is equally enmeshed in tempest-like feelings for Tapia. In 1857 he uses the aftereffects of a life of storms to summarize his sad state, exiled in Cuba, when he compares himself to a battered palm tree, "como yo cansada de resistir al viento de las tempestades y las desdichas ha concluido por abandonar su ramaje, indiferente, a los empujes del viento veleidoso" ("[the palm] like me, is tired of resisting the wind of storms and misfortunes, has finished by abandoning its branches, indifferent to the thrusts of the fickle wind"; 1968, 580). In *La palma del cacique* ("The Cacique's Palm"), Tapia paints the rage of the hurricane as the only proper setting for exile, "[s]i durante la noche el viento brama, el relámpago brilla, retumba el trueno, la lluvia cae a torrentes, se oyen de vez en cuando los acentos de un hombre, que llora su país natal" ("if during the night the wind roars, lightning flashes, thunder rumbles, rain falls in torrents, the expressions of man are heard from time to time, crying for his native country"; 1968, 205). The protagonist in this environment is still Man against Nature; the metaphorical force is exile, a common topic among nineteenth-century Puerto Rican writers, many of whom, including Tapia, were exiled for political reasons.

José de Diego's "En la brecha" ("Unto the Breach," 1900) is exemplary in the uses of "storm" as a metaphor for the slings and arrows of the cruel fate of a colony. A bull braving San Ciriaco—plus the metaphorical storm that was the new U.S. colonialism—became a gripping symbol of national resistance, a fiercely *independentista* call for courage against colonial forces who oppose the ideal of independence. School children have had to read:

> De la tormenta el iracundo empuje, / No has de balar como el cordero triste,
> sino rugir, como la fiera ruge! / ¡Levántate, Revuélvete, Resiste! / ¡Haz como el toro acorralado: muge! / ¡O como el toro que no muge embiste! ("At the storm's enraged attack, / you will not bleat, like a sad lamb, / but roar, like the wild beast roars. / Arise! Seethe! Resist! / Do as the bull

> at bay: Bellow! / Or like the bull that doesn't bellow: Lunge!" (1904, 168).

The bull, a potent animal of symbolic strength, is an emblematic call to political activism. The ecocritical context enhances the poem's force: it appeared one year after San Ciriaco, one of the major hurricanes of the time, a villainous character whose name is still recognized, having traumatized Puerto Rico with 3369 deaths and untold damage.

The pendulum from ecoerasure to ecophilia undergirded by fierce political idealism will become customary in Puerto Rican literature. Fierce anarchist labor leader Luisa Capetillo writes in 1916 of defying the storm, asserting human dominance over a fierce Nature: "La tormenta es poderosa, y yo la he desafiado. El abismo es tenebroso, y yo no lo he temido. El mar es peligroso y yo lo he cruzado" ("The storm is powerful, and I have defied it. The abyss is dark, and I have not feared it. The sea is dangerous and I have crossed it" 2008, 82).

The garden allegory coexists with ecoerasure in the late nineteenth century. Lola Rodríguez de Tió, a poet in the late Romantic mode, strategically uses both the garden trope as well as storms. Her poem "A Puerto Rico," for example, begins firmly aligned with the mild, flowery image of "Borinquen, nido de flores" ("Borinquen, nest of flowers"). Yet in her book *Claros y nieblas*(from 1885, during the gap in hurricanes between 1876 and 1888) she admires the North Sea in the sublime Romantic mode for which a stormy sea is emotionally stirring: "el estruendo de las notas / con que expresas las tormentas" ("the din of the notes / with which you express the storms"; 1885, 8), she depicts a roaring seascape while still preserving the placid trope of lapping waves caressing the land. The book is full of stock images of turbulent seas mirroring the poet's moods (love, nostalgia, motherhood, exile, and other topics). Yet, despite notes on oceanic power, Rodríguez de Tió mostly follows the tradition of flowery portraits of an Edenic landscape, in sharp contrast to the emotional hurricane of her combative political poem "La borinqueña."

The Edenic isle allegory that was firmly established early on persists until late twentieth century. Lilliana Ramos Collado summarizes the depth and extent of this traditional view:

> Presente en nuestra literatura desde que los primeros cancioneros que le cantaban a nuestra bella isla doncella, y enredado en el afán de nuestros primeros novelistas naturalistas que querían curar los males de la patria

para devolverle su salud y belleza virginales, el Edén puertorriqueño sigue vivito y coleando en la "Alabanza en la Torre de Ciales" de Juan Antonio Corretjer y en la empresa educativa de la DIVEDCO durante la década de 1950, en los paisajes espléndidos de Myrna Báez en los 70s y 80s, en nuestra música neojíbara o neocriollista ("Present in our literature since the first songbooks that sang to our beautiful maiden island, and entangled in the zeal of our first naturalist novelists who wanted to cure the ills of the homeland to restore its virginal health and beauty, the Puerto Rican Eden is still alive and kicking in the "Praise in the Tower of Ciales" by Juan Antonio Corretjer and in the educational enterprise of DIVEDCO during the 1950s, in the splendid landscapes of Myrna Báez in the 70s and 80s, in our *neojíbara* or *neocriollista* music"; 2012)

Yes to all that, except she was wrong about DIVEDCO, as we will see in the next chapter.

The Serene Landscape in Art

With its placid landscapes Puerto Rican painting defies Romantic trends. European Romantics made the landscape superlatively important in art, elevating the genre from its earlier lowly status. Romantic art liberated the painterly landscape from the low ranking it had in the seventeenth-century French Academy, which had been vastly influential in the art of the Americas.[6] Landscapes became "the finest achievement of Romantic painting" (Janson 1977, 470) as Nature assumed protagonism in artistic expressions of nineteenth-century Western art. Realism later produced many pristine landscapes, and Impressionism (a movement to which Francisco Oller belonged), elevated landscapes even more in its search for connections to visual moments in time. Caspar David Friedrick's Rückenfigur man in *The Wanderer* dramatically looks down on clouds and mountain peaks. European Romantic painting also leads to "the recurrence of images of shipwrecks... and other representations of man's struggle against the awesome power of nature" (Calley Galitz N.d.).[7] Mid-nineteenth century the American Hudson River School displayed the grandeur of sweeping U.S. landscapes.

The main Puerto Rican artists of the eighteenth and nineteenth centuries show a progression from a Europeanist mode of portraiture to an increasing focus on local panoramas. José Campeche, with his abundant portraits of important people, uses Rococo formulas of sumptuous interiors and pulls back curtains to the panoramas outside. His

Nature is mostly bucolic scenery in the distance, yet increasingly he and other Puerto Rican visual artists portray specific local flora with symbolic ties to national identity. Campeche uses pineapples at least twice in his portraits,[8] and also other indicators of location. We also see non-humans that are useful to humans and important to the nation's economy: produce to eat and fauna such as the pig to consume (in Oller's *El velorio*), or a human using an ox or riding a horse.

Local produce is fertile ground for patriotic love of the homeland, *la patria*, and is crucial for the development of a national culture. Campeche progressively constructs a national identity through increasing depictions of regional flora. His Rococo-style paintings transition from an interior to an exterior environment as the landscape in the background grows more and more specifically Puerto Rican. For instance, his *Dama a caballo*,[9] from 1785 (the year of hurricane San Lupo) displays a traditional approach to generic European nature as a background to the human protagonists in the foreground. In contrast, Campeche's 1790 dramatic narrative painting *The Rescue of Don Ramón Power y Giral*[10] depicts a tempestuous oceanic environment. It portrays the rescue adventure of the title, as the young boy is taken from the storm-blown waves of the sea into a small boat. That boy, Ramón Power y Giralt, future leader, stares at the viewer impassively, as if to say no storm will defeat him (he is a hero after all). Alas, for our ecocritical purposes Campeche, an artist still very tied to European painterly themes, did not represent in this painting a local or Caribbean background for this heroic adventure, because the vessel La Esperanza sank off the coast of Cantabria.

Campeche progressively opens the curtain on local panoramas, yet rarely depicting disasters. Partly the indoors bent is due to his training in European portraiture. Many of the portraits for which he is famous are in interiors, for instance *Retrato de dama* (c. 1792). A year earlier, though, an unnamed hurricane had blown through the region. Campeche comes closer to a local environment in *Don José Mas Ferrer* (c. 1795) with a Caribbean bay and setting sun sky seen through a window. By the time of *Gobernador Don Ramón de Castro*[11] (c. 1800) Campeche depicts a recognizably local landscape; including views of a beach and sea. He was not necessarily averse to human violence seen from a distance; in *Gobernador Don Ramón de Castro* the subject points to a battle scene raging in San Juan Bay, overseen by the looming walls of El Morro. In Campeche's portraits of great men and women the gravitas and drama of important people reflect a prosperous people in control of a placid environment.

These prominent people are closer to the mood we saw in early modern stories featuring heroes conquering the land (Fig. 3.1).

In 1792 Campeche's *El Gobernador Don Miguel Antonio de Ustáriz* repeats the formula of pulling back a curtain but here it reveals San Juan's infrastructure under construction. This important painting shows a dense cityscape, and in the far background a bay and mountains. This may actually be the first Puerto Rican painterly depiction of infrastructure, a topic that rages in Puerto Rican culture from late nineteenth century into the present day. Here workers lay down city cobblestones as citizens look on with interest; community life under the important Man who holds the blueprints to the Nation.

Edgardo Rodríguez-Juliá labels Campeche a painter of interiors with a fondness for "el espacio cerrado" ("the enclosed space"; 1986, 10) and says the Ustáriz portrait illustrates how "la pintura de Campeche se distancia de la naturaleza. En Campeche no advertimos esa curiosidad por la campiña tan característica del Goya rococó" ("Campeche's art distances itself from nature. In Campeche we do not perceive that curiosity for the countryside so characteristic of the Rococo Goya"; 1986 11). A painter of interiors, certainly, but also, as Rodríguez-Juliá points out, a painter of State and power. And perhaps that is why infrastructure seeps into this portrait, because these new heroes are Great Men who build nations.

One of Campeche's religious paintings is the exception to the near-universal lack of representations of disaster in his time. As we saw in Chapter 2, Campeche, who painted many works with religious themes, produced *San Emigdio de Áscoli* in 1787, the same year that an earthquake struck the island. In this painting of the patron saint of earthquakes, Campeche depicts buildings crumbling. Even though the painting fits the religious subgenre, it is still a remarkable depiction of a natural disaster at a time when none appeared in Puerto Rican art. It is the exception that proves the rule.

Francisco Oller y Cestero, a momentous painter, fully engaged with a placid local environment. His entanglements with his local environs set the templates for many depictions of nature still in use. His depictions of the Puerto Rican landscape are absolutely foundational to Puerto Rico's landscape art. In Chapter 1 we discussed how the sublime skies of *Hacienda Aurora* underscore racialized exploitation. The Museo de Arte de Puerto Rico summarizes his towering importance for representing authentically native vegetation: "Inició en la isla el interés por llevar al lienzo la realidad puertorriqueña, captando la belleza del paisaje, nuestros

Fig. 3.1 José Campeche y Jordán. *El Gobernador Don Miguel Antonio de Ustáriz*. Oil on wood, 1789–92. Instituto de Cultura Puertorriqueña Collection

frutos tropicales y aspectos de nuestra problemática social" ("He began the interest on the island for putting Puerto Rican reality on canvas, capturing the beauty of the landscape, our tropical fruits and aspects of our social problems").[12]Oller's local flora resonates with national symbolism, portraying in 1887–1888 another tree symbolic of national strength, *La Ceiba de Ponce* ("The Ceiba Tree of Ponce").[13] It is the oldest of his Puerto Rican landscapes and centers the majestic national tree. His *bodegones* (still life paintings) feature local produce and plants in profusion, such as *Piñas* ("Pineapples") from 1890 (Fig. 3.2),

Similarly, *Bodegón con guanábanas* ("Still Life with Soursops")[14] from 1891 strengthens his depictions of local produce. 1910's *Paisaje con palma real* ("Landscape with Royal Palm")[15] and 1912's *Paisaje de Puerto Rico* ("Landscape of Puerto Rico")[16] feature the iconic *palma real* along with other native plants, therefore depicting an authentic Puerto Rican landscape, as opposed to the European environment that had been rendered so often by his predecessors.

He devotes many of his works to exalting local vegetation and landscapes including some in his *magnum opus*, *El velorio* ("The Wake" in 1893).[17] *El velorio*'s main subject is the nation's people, not Nature, and

Fig. 3.2 Francisco Oller y Cestero. *Piñas*. Oil on wood, c. 1890. Instituto de Cultura Puertorriqueña Collection

it represents the nation in all its gendered, religious, racial, and hierarchical complexity. Apart from its famous *racimo de plátanos*, it gazes at a gentle Nature through windows and doors to show people at work in a placid impressionist backdrop. As Antonio Martorell observes, the *racimo* had been a protagonist in *El velorio*, elevated even higher than the pig, and, most importantly for us, it brought the landscape inside: "mi verde hace entrar el paisaje en la humilde habitación" ("my green brings the landscape into the humble room"; 2010, 47).

Yet how striking that his ecoerasing portrayal of verdant, sweet nature outside the modest house that is the setting for the painting was created after 1889's hurricane San Moisés and 1891's San Magín. It was finished the same year San Roque II caused, among deaths and damages, the following: "The coffee and minor fruit crops were lost at a cost of 150,000 pesos. Most of the poor people's houses were blown down" (Boose). Yet *El velorio* is rooted in a traditional landscape untroubled by storms.

The Instituto de Cultura Puertorriqueña's (ICP) database as well in their book *Pintura puertorriqueña* offer a plethora of landscapes and cityscapes from all eras of Puerto Rican painting. Among the many subjects covered in this vast collection are: religion, history (particularly important historical figures), colonization, politics (including many portraits of political movers and shakers), still life paintings of local plants (including some of Oller's *bodegones*), abstract art, folkloric traditions, everyday working folk, cityscapes, and finally, many, many landscapes. Most of these feature vegetation closely associated with the Nation, reflecting ties to the "authentic" Land (among them the beloved palm and *flamboyán* trees). For instance, Miguel Pou Becerra's *Las calesas de Ponce* (N.d., 86) from 1940 features the iconic *flamboyanes* in the foreground and the cathedral in the background linking a placid, iconic symbol of nature with religion as elements central to Puerto Rican life.

As the ICP collection shows, Puerto Rico's artistic tradition focuses on a wide variety of markers of national identity. The ICP's online exhibit "Icons of National Identity,"[18] is *costumbrista* in its focus and prominently features iconic *jíbaros/as* because starting in the late nineteenth-century artists highlight working-class people, especially those closest to the land. The permanent collection has Ramón Frade's iconic *El pan nuestro*, (CPE. 1991. 0282) which shows a shoeless *jíbaro* with his *pava* hat carrying a plantain *racimo*. Behind him is a vista of the countryside, including a palm tree. The painting appeared in 1905, at the beginning of the U.S. takeover of the island and just six years after San Ciriaco. In 1899 when San Ciriaco passed over Adjuntas—prime *jíbaro* countryside—"the crops were destroyed" (Boose). In that ecological context Frade is paying tribute to the resilience of the vegetation as well as the people. Frade's *jíbaro* and the *racimo* are one and the same: easily recognizable and dependable representations of Puerto Rico, authenticity in human and plant form. As the ICP's database demonstrates, considerable artistic effort was expended throughout the twentieth century on placid landscapes that are center stage, with no works depicting hurricanes nor their aftermath.

Nevertheless, storms were the province of some notable global artists in the nineteenth century, so we can compare Puerto Rican artists in the nineteenth century to disaster art produced elsewhere. Landscapes everywhere tend to depict gentle natural environments yet a few great artists exploit storm drama, and when storms appear, they dazzle. Exhibit A is Katsushika Hokusai's *The Great Wave off Kanagawa*,[19] a world classic that depicts a tsunami[20] in the series *Thirty-Six Views of Mount Fuji (Fugaku sanjūrokkei)*, and is from 1826–36 (Fig. 3.3).

Winslow Homer, one of the United States' most storied and still massively popular artists, was a prolific painter of seascapes, especially stormy seas. He painted many dramatic seascapes, among them *The Life Line* (1884),[21] a rescue at sea. Homer also painted racially-conscious depictions of people battling storms, most famously *The Gulf Stream* (1899),[22] a dynamic action painting depicting a Black man in a tattered little boat about to capsize. The sea is stormy and there is a shark underneath—nature at its most dangerous—yet the work posits a possible happy ending of a rescue from a distant boat. In the 1899 watercolor *After the Hurricane, Bahamas*[23] the Black man is shown washed up on shore, battered but alive. In both works a Black man, tellingly, faces the threat of the storm alone. There is a dramatic depiction of a rescue at sea in José Campeche's *The Rescue of Don Ramón Power y Giralt* in which people work together to save a child but alas, set in the waters of Cantabria.

Homer also painted rare images of women battling oceanic elements. *The Gale* (1883–93)[24] presents a mother on the shore with a baby (who looks at the viewer) strapped to her back as they are buffeted by stormy winds. *Inside the Bar* (1883)[25] also features a North Sea fisherwoman carrying a basket as storm winds batter her. The Metropolitan's commentary states Homer presents "heroicized figures engaged in dramatic action"; the heroes here are women (Fig. 3.4).

Fig. 3.3 Katsushika Hokusai. *The Great Wave off Kanagawa*, from the series *Thirty-Six Views of Mount Fuji (Fugaku sanjūrokkei)*, color woodblock print; oban, 1826–36. The Art Institute of Chicago

For our ecocritical purposes, though, nothing tops Winslow Homer's *Hurricane, Bahamas* (1898) which depicts houses and palms beaten by a hurricane. It is a rare nineteenth-century depiction of a specifically Caribbean cyclone. As the Metropolitan Museum says, "[a]s was his tendency, the artist acknowledged the archipelago's status as a Crown colony—signaled here at bottom center with the Red Ensign, a flag flown by British ships. This detail combined with the tempestuous weather may evoke the geopolitical turmoil elsewhere in the Caribbean that year, specifically in Cuba and Puerto Rico."[26]Homer accomplishes what few other artists of the time did, commenting on the twin storms of imperial control and hurricanes in the Caribbean. Such is the staying power of this work that is the book cover of Dorsía Smith Silva's 2024 post-Maria book of poetry, *Inheritance of Drowning*.

Fig. 3.4 Winslow Homer. *Hurricane, Bahamas*, watercolor, 1898. The Metropolitan Museum of Art. Public domain

Late Nineteenth-Century History: Selective Ecoerasure

Politics are at the crux of the effects of hurricanes on Puerto Rico's financial situation, in circumstances that appear in uniform manner from 1530 to Salvador Brau.[27] The island's political dependency and the ever-present fact of diaspora are of equal importance regarding the impact of environmental disasters. Brau's *Puerto Rico y su historia: investigaciones críticas* ("Puerto Rico and Its History: Critical Research") from 1894 opens on Ponce de León's hazardous adventures, but without mentioning his deeds during San Roque, which Ponce de León recorded as the first hurricane to hit the island, nor the three hurricanes of 1530. Brau explains that the word hurricane came from the Taino language, noting its religious overtone of an "espíritu maligno" ("malign spirit"; 1894, 19) but he mentions storms as an aside after describing an assault by "guerreros indios" ("Indian warriors") in Cristobal de Guzmán's farm (an event we saw in Chapter 2).[28] Brau forgets the woman in that story, as told by Abbad (see Chapter 2). Brau describes violent weather as a common Caribbean trait: "Son comunes en las Antillas durante el periodo canicular esos fenómenos meteorológicos que llevan hoy el nombre de 'ciclones' y á que los indios llamaban 'huracanes'" ("These meteorological phenomena that today bear the name of 'cyclones' are common in the Antilles during the canicular period, and which the indigenous people called 'hurricanes'"; 1894, 246). Yet his retelling of Puerto Rican history largely elides hurricanes, a Caribbean metanarrative.

Puerto Rico's need for aid and the heightened migration after storms were as critical in 1530 as they were after hurricane Maria in 2017. Brau outlines the colony's financial situation in 1530 as the increasingly slave-dependent colony sought redress for storm damages from the metropolis. He specifically correlates finances to hurricanes when he places his account of the storm in a section on governance titled "Gobernación y gobernadores" ("Governing and Governors"). There he describes as "interesting news" a letter Governor Manuel de Ovando sent to the authorities in Spain that describes the colony's sorry state post-hurricane, begging for relief (1894, 123). Brau ties the 1530 storms to tax collecting, "fue tal su violencia y tal grado de miseria produjeron que uno en especial es nombrado para activar la recaudación de tributos, debido a la general penuria" ("such was their violence and such a degree of misery they produced that one [government official] is specially appointed to start collecting tributes, due to the general penury"; 1917, 66–67). He notes that in 1537 three devastating hurricanes caused flooding, damaging "labranzas, ganados y esclavos" ("farms, cattle and slaves"; 1917, 78), in that order.

Brau's ecoerasures are symptomatic of how the ecoerasing "forgetting" of storms worked in Puerto Rican culture despite the realities of the ecoregion. Born in Cabo Rojo, a coastal area prone to hurricane devastation, he was on the island for hurricanes in 1848, 1850, 1851 (a horrible year with a hurricane and earthquakes), and 1852 (which also had a yellow fever epidemic and a hurricane). He was in Spain during 1867's San Narciso and its earthquakes but he returned to Puerto Rico in 1870 and so he was there for Santa Juana in 1871 and 1876's San Felipe I (not the massive San Felipe II), as well as 1888's San Gil, 1893's San Roque II, 1889's San Moisés. Back from Spain where he resided from 1894–1897, he was the island's appointed official historian until his death in 1912 so he was present for 1899's earthquakes and monster San Ciriaco (Category 4). Given this personal experience of hurricanes it is striking that though Brau sometimes notes hurricane destruction, they are not central to his historical viewpoint, whereas other things are; he spends ten pages of *Puerto Rico y su historia* ("Puerto Rico and its history") analyzing the spelling of Agüeybaná's name in the Spanish language (1894, 55–66). The book was published in 1892, one year after hurricane San Magín caused 8 deaths and extensive flooding. By contrast, in *Historia de Puerto Rico* from 1904, he devotes a section to San Ciriaco in 1899, lauding the help the United States and the government gave (308). The book also has a sub-heading "huracanes y corsarios" ("hurricanes and corsairs") which mentions the early colonial devastations of two hurricanes alongside what he considers the even more disastrous "plague" of corsairs (1904, 227).

Ecoerasing representations of the Land provide a framework for one of the most contentious debates in Puerto Rican culture, docility as a national character trait. A mythically serene environment encourages the view that people are pacific or, infamously docile, through the process of the mirroring of a fictionally mild land and equally fictional docile people. Brau, for instance, endorses the view that geography is destiny. Looking to pre-Columbian times, his essentialist view of the bioregion's indigenous inhabitants is evident:

> La raza caribe, esparcida por el vasto territorio que se extiende desde el Ecuador hasta las Antillas, distinguíase por su espíritu belicoso y emprendedor. Eran estas cualidades generales que en Boriquén . . . se transformaron en sentido opuesto, distinguiéndose los naturales de la primera por sus condiciones hospitalarias y generosas, por sus instintos sedentarios, su docilidad, el apego al terruño y las aficiones agrícolas. ("The Carib race, scattered across the vast territory that stretches from Ecuador to the Antilles, was distinguished by its warlike and enterprising spirit. It was these general qualities that in Boriquén . . . were turned in the opposite direction, the natives of the former being distinguished by their hospitable and generous conditions, by their sedentary instincts, their docility, attachment to the land and agricultural hobbies"; 1894, 77).

Brau believes a placid land makes people timid and later, he extols how even a people he regards as inferior are also transformed by this placid Nature: "La rudeza salvaje del etíope se suaviza" ("The wild brusqueness of the Ethiopian softens"; 1894, 78).

Brau subscribes to a view of a benevolent Nature that overflows with sweet abundance and believes his impossibly placid environment molds people into timidity and submissiveness:

> Entre la grama lujuriosa de nuestros prados no se esconde la víbora artera ni rastrea el crótalo ponzoñoso . . . no surcan en las corrientes de nuestros mansos riachuelos ni el aligator monstruoso . . . No interrumpen la correcta curva de nuestras suaves colinas, la cortadura de Rocallosas Cima ni el cono de pavoroso volcán; extensa alfombra de verdura, mantenida por perpetua primavera . . . el mar lame mansamente nuestras playas . . . Todo parece conspirar á la mansedumbre y el sosiego en esta tierra privilegiada . . . ("The cunning snake does not hide among the luxurious grass of our meadows nor does the poisonous rattlesnake slither . . . in the currents of our meek streams the monstrous alligator does not swim . . . Nor do the jagged summits of the Rockies nor the cone of a terrifying volcano interrupt the precise curve of our soft hills; extensive vegetable carpet, maintained by perpetual spring . . . the sea meekly skims our beaches . . . Everything seems to conspire to meekness and tranquility in this privileged land; 1894, 78).

We will return to docility in the next chapter but it is clear Brau ties a placid land to patriarchal, heteronormative romantic and familial ties, always within a strict gender binary. He reaffirms the values of a patriarchal family structure and its attendant forms of nationalism which did not extend, in Brau's case, to political independence, the fight for which had been fierce in his times:

> Tierra creada para el amor . . . mas no para el amor volcánico, tempestuoso, . . . sino para ese afecto sosegado y tierno que se amparan las intimidades de la familia, y acendra virtudes anodinas del hogar y del terruño, y eleva las angostas funciones de la maternidad hasta los transportes fanáticos de la idolatría ("A land created for love . . . but not for volcanic, tempestuous love. . . but for that calm and tender affection that is protected by the intimacies of the family, and will accrue bland virtues of home and homeland, and lifts the narrow functions of motherhood to fanatical transports of idolatry"; 1894, 79–80).

Therefore, Brau illustrates the deep links between the bioregion's imaginary serenity and a heteronormative and essentialist picture of humans with equally submissive women to fit the landscape. He erases human sexuality like he does hurricanes and earthquakes, in order to exalt the heteronormative virtues of his time: chaste love and motherhood.

Manuel Zeno Gandía's novel *La charca* ("The Pond") was published in 1894 after a series of hurricanes in 1889, 1891, and 1893. As a naturalist novel, it offers a grim view of people, as shown by the number of murders and assaults in its plot. Zeno Gandía diagnoses the illness of colonialism as the main metaphor for life in his unequivocally titled series *Crónicas de un mundo enfermo* ("Chronicles of a Sick World"). He stresses illnesses caused by colonial underdevelopment and the flawed nature of humans (alcoholism, lust, greed). His enduring figuration of the precariousness of country life in Puerto Rico features mostly man-made disasters (all sorts of violence, from rape to murder) all exacerbated by colonialism tied to land ownership. His spokesmen in the novel, Dr. Pintado and his colleague don Juan del Salto, have a science-oriented, positivist view of life, regarded as the only hope to cure what ails the nation.

For the most part *La charca* uses Nature and storm forces metaphorically, for instance, human life is a torrent leading to death: "Horrible corriente, que va fatalmente a la muerte! ¡Caudal de vida condenado a extinguirse bajo la depresión constante que fermenta en los organismos!" ("Horrible current, which goes fatally to death! Life's flow doomed to extinction under the constant depression that ferments in organisms!" N.d.).

It features, however, one powerful exception to depictions of a placid Nature: a dynamic tale of a rescue in an overflowing, stormy river during "the day of the great rainfall" which he does not, alas, tie to any specific hurricane. Zeno Gandía characterizes the overflowing river as a monster that matches the violence of men: "[e]l torrente precipitábase en una carrera sin freno, aullando como can enfurecido, retorciéndose como gigantesca serpiente, resuelto a romper la estrechez del canal que lo encauzaba" ("the torrent rushed in an unbridled race, howling like an enraged dog, writhing like a gigantic serpent, determined to break through the narrowness of the channel that guided it," N.d.). For one moment human emotions and violence (Marcelo's doubts and Montesa's abuse of Marcante) all recede before the howling wind, blood-like furious waters, and mudslides. Zeno Gandía's flood waters bring violent metaphors: "El aire se estremecía, invadido por el estrépito, y sus sacudimientos bufaban como si en aquel momento descargara un odio secular" ("The air shuddered, filled with the din of it, and its jerks snorted as if at that moment it vented a centuries-old hatred," N.d.), and "rugían, en fin, con ira de chacal encadenado" ("they roared, in short, with the wrath of a chained-up jackal," N.d.), and "aquel canjillón iracundo por donde corría la muerte ("that ditch of wrathful mud through which death flowed," N.d.).

And yet in the moment when the waters threaten to drown a boy, seven men heroically band together and rescue the child: "[e]l monstruo líquido tuvo que romperse para dejar penetrar en su seno a algunos jirones de Humanidad ennoblecidos por la grandeza de los heroes" ("the liquid monster had to break up to allow some shreds of Humanity, ennobled by the greatness of heroes, to penetrate into its center," N.d.). It is one of the few moments of joy in a novel full of people mired in unhappiness and violence, as a unified group of men triumph over Nature.

Zeno Gandía, however, has a dual vision of the landscape, using both the savage event of mountain floodwaters as well as traditional springtime tropes that open and close the novel. It opens on a fantasy of a garden of eternal spring like many in the nineteenth century:

> con aquel reguero de colores eran los campos la inmensa paleta en donde había de humedecer sus pinceles el supremo artista. Un azul inimitable descendía del cielo como regalo nupcial, y un verde suave parpadeaba en las campiñas como ofrenda esclava. . . . Los árboles, en eterna gemación, ostentaban vestiduras rosadas y galas rojas, y así mostrábanse los paisajes como proyectados al mundo de los sueños por la mano de la primavera ("with that trail of colors the fields were an immense palette on which the supreme artist moistened his brushes. An inimitable blue descended from heaven as a nuptial gift, and a soft green flickered in the fields as a slave offering. . . . The trees, eternally budding, wore pink robes and red finery, and thus the landscapes were shown as if projected into the world of dreams by the hand of spring", N.d.).

At the end of the novel a mild Nature is a witness to, not the cause of, human suffering. Silvina's tragic ending is brought on by abuse and illness, yet as she dies, the gentle sounds of a placid nature surround her:

> Cuando los últimos clamores de la desgracia desvaneciéronse a lo lejos, el lugar quedó solitario. Sólo el río quedó murmurando inquieto, siempre sonante, como si arrastrara en su corriente el prolongado lamento de un dolor sin bálsamo, como si llevara disuelto en su linfa el llanto de una desdicha que nadie enjuga, que nadie consuela, ¡que nadie conoce! ("When the last cries of misfortune faded away, the place was desolate. Only the river murmured restlessly, always resounding, as if it dragged in its current the prolonged lament of suffering without balm, as if it had dissolved in its waters the cry of a misfortune that no one cleanses, that no one comforts, that no one knows!", N.d.).

As Nature's counterpoint to the stagnant pond of the title, the babbling brook becomes the sole witness to the misery and woe caused by humanity's destructive impulses.

Lesser-known writer Jose Pérez Losada, a Spanish emigré, refers to San Ciriaco in his two-part Naturalist novel, *La Patulea* (1906) and *El manglar* (1908). Miguel Angel Náter, director of the Seminario Federico de Onís at the University of Puerto Rico, Río Piedras, says these two early twentieth-century novels "include the passage of Hurricane San Ciriaco through Puerto Rico as an expression of the deterioration caused by the invasion of 1898."[29] As Manuel Gómez Rivera says, these novels have a core independentista message (1969, 51) and so the symbolic value of personifying colonialism as a hurricane allows for Losada's political hyperbole: "todo caía ó amenazaba ruina en el brutal sacudimiento de la catástrofe. Todo, leyes y costumbres personalidad y caracteres, instituciones y organismos, seres y cosas, todo barrido en la racha del huracán" ("everything falling or threatening ruin in the brutal tremor of catastrophe. Everything, laws and customs, personality and characters, institutions and organisms, beings and things, all swept away in the hurricane blast") (1908, 288).

Despite these few naturalist examples, exceptionalist portrayals of a serene environment dominate Puerto Rico's cultural production throughout the nineteenth and early twentieth centuries. These are the same spectacles in which writers exalt what Miguel Meléndez Muñoz will call, in 1936, "las maravillas de nuestra isla" ("the marvels of our island" 6) with little regard for the dreaded threats of the bioregion.

Of course, government and commerce take advantage of the fiction of *La Isla del Encanto* ("The Isle of Enchantment"). This touristy catchphrase is traditionally associated with a travel guide (of the same name, from 1938) written by Theodore Roosevelt Jr. The irony is that it had only been six years since San Ciprián: Category 4, 225 deaths in P.R. and U.S. Virgin Islands and $657,893 in damages, adjusted for inflation (NOAA *Deadliest*). The catchphrase goes back even further; AGWI Steamship Lines touted the ease of access to "Porto Rico" labeled as the "Island of Enchantment" and reachable from New York in only four and a half days in 1913 in a magazine called *The Judge* and the slogan reappeared in 1914 in *House and Garden Magazine*. These ads both correspond to the gap in hurricanes between 1899 and 1916, which does not, of course, excuse the chicanery of selling the fiction of an endlessly mild climate. The slogan has had longevity and popularity, as we see in the thousands of followers of the nostalgia-tinged Facebook group *Isla del Encanto*.

From the nineteenth century into the twentieth century most historical, literary, and artistic creators ecoerase hurricanes and earthquakes. We have explored many of the reasons fueling this culture of ecoerasure. Gaps in hurricane activity are a reason to forget or at least gloss over disaster. Literature and art also adapted Romantic descriptions of Nature to go along with their preferred placid landscape. This will all change in the twentieth century.

Notes

1. In a story that was not included in his either of two collections published in 1818 and 1860.
2. Always capitalized in the Spanish of the times, a graphic sign of exceptionalism.
3. Although the barometer started providing some early warning in 1837.
4. A style focused on customs and social manners.
5. Not 1928's monstrous San Felipe-Okeechobee.
6. An attitude not present in the splendid and unconventional art of the Netherlands at the time.
7. Campeche follows this trend with *El salvamento de Don Ramón Power* (c. 1790).
8. In *Dama a caballo* (1785) and *Las hijas del gobernador Ramón de Castro* (1797).
9. https://web.archive.org/web/20250210160828/https://artsandculture.google.com/asset/lady-on-horseback-jos%C3%A9-campeche-y-jord%C3%A1n/dwFVWJOw_LfxuA
10. https://web.archive.org/web/20250210161131/https://commons.wikimedia.org/wiki/File:Rescue_of_Ramon_Power.jpg
11. https://web.archive.org/web/20250212181758/https://www.mapr.org/en/art/artwork/gobernador-don-ramon-de-castro
12. https://web.archive.org/web/20250210162409/https://mapr.org/es/arte/obra/estudio-para-el-velorio
13. https://web.archive.org/web/20250210164323/https://z-upload.facebook.com/museoarteponce/posts/4364087956973626
14. https://web.archive.org/web/20250210162850/https://www.mapr.org/es/arte/obra/bodegon-con-guanabanas

15. https://web.archive.org/web/20250210163342/https://www.mapr.org/es/arte/obra/paisaje-con-palma-real
16. https://web.archive.org/web/20250210163904/https://artsandculture.google.com/asset/paisaje-de-puerto-rico-francisco-oller-y-cestero/OAHdONh5LdGiJg
17. https://web.archive.org/web/20250210164430/https://www.uprrp.edu/tag/museo-de-arte-upr/
18. https://web.archive.org/web/20250212180358/https://artsandculture.google.com/story/mwVhhwzusbr6LA
19. https://web.archive.org/web/20250210164728/https://www.artic.edu/artworks/24645/under-the-wave-off-kanagawa-kanagawa-oki-nami-ura-also-known-as-the-great-wave-from-the-series-thirty-six-views-of-mount-fuji-fugaku-sanj%E7%AC%9Brokkei
20. The Pacific Coastal and Marine Science Center states that, "While the real intent and meaning behind Hokusai's *Great Wave* woodblock print is not known, and while tsunami waves typically do not crest or break and do not have white caps like this, Hokusai's print is nonetheless often used to depict a tsunami."
21. https://web.archive.org/web/20250212181109/https://artsandculture.google.com/asset/the-life-line/xQE9eAXWkRGPUQ
22. https://web.archive.org/web/20250212181438/https://artsandculture.google.com/asset/the-gulf-stream-winslow-homer/gAEvDCycRCZdlg
23. https://web.archive.org/web/20250210164835/https://www.metmuseum.org/art/collection/search/829793
24. https://web.archive.org/web/20250210165123/https://www.metmuseum.org/art/collection/search/837835
25. https://web.archive.org/web/20250212174624/https://www.metmuseum.org/art/collection/search/11125
26. https://web.archive.org/web/20250210165621/https://www.metmuseum.org/art/collection/search/11124
27. Brau was named Official Historian of Puerto Rico in 1903 by the American-appointed governor William Henry Hunt.
28. Brau forgets the woman in that story, as told by Abbad (See Chapter 2).
29. In email 12 Feb. 2024.

CHAPTER 4

Climate Anxiety, Trauma, and Infrastructure in the Twentieth Century

It is a truism that in Puerto Rico they paved paradise, more ironic when we consider that Puerto Rico's nineteenth-century culture closely identified national identity with a pristine, unspoiled landscape. Creators in the twentieth century then moved from myths of a timeless, serene landscape to depictions of the ecoreality of an often-brutal environment. This chapter traces twentieth-century storytelling and artistic routes that interpret the environment, especially disasters—both environmental and stemming from colonialism and capitalism. These catastrophes lead to climate anxiety and despair, yes, but also offer proof of the grit and resilience of Puerto Rican communities.

In Chapter 3 we saw Puerto Rican culture awash in images, symbols and stories that ecoerased hurricanes and other disasters from the beloved Land. These sanitized images of an *Isla* of eternal springtime plummet after 1899's San Ciriaco and 1929's San Felipe. In highbrow culture[1] bucolic landscapes become increasingly unsustainable in the twentieth century due to the impact of humans on the landscape, particularly fossil fuel infrastructures (roads and the electrical grid). Advances in predicting and tracking hurricanes—from differential equations to hurricane hunter airplanes—also impacted art, as it became easier to "see" hurricanes.

Increasingly, Puerto Rican culture create products that offer, as Stephen Daniels and Hayden Lorimer say, "no immediate pastoral comfort, little in the way of consolation, and that instead document

M. Acosta Cruz, *An Ecocritical Study of Puerto Rican Culture*,
https://doi.org/10.1007/978-3-032-04044-2_4

the dread of a felt world in a time of environmental reckoning" (2012, 5). By at least mid-century, writers and artists engage facets of ecology, geology, and meteorology in ways that reflect material reality. Hurricanes again become dynamic metaphors and characters, as they had been in early modern times; agents in the story of the Nation. For example, this cultural shift is signaled by the founding in 1975 of Ediciones Huracán, a distinguished publishing house. Late in the century ecoerasure becomes implausible as climate change warms ocean waters and intensifies the possibilities of supercharged storms (Cheng, 2022, 777).

A toxic colonial mix of federal and local inefficiencies always impact infrastructures and disaster management. Since Chapter 2 we have seen the connection between disasters, government, and infrastructure; witness Juan de Vadillo's 1527 letter to the emperor (Tapia 1854, 295). Imposed U.S. control is regularly deficient and exploitative, proof is the Jones Act,[2] as well as ineffectual U.S. disaster agencies. For example: "The New Deal's patchwork of interventions there —*hurricane relief*, educational mandates, agricultural re-organization— created more turmoil than succor" (Schecter 2015, 158, our emphasis). Puerto Rico's burdensome territorial status undergirds representations of the environment throughout the twentieth century.

In the twentieth century Puerto Rican historical writing once again features hurricanes and earthquakes as it did in the early chroniclers. Historical writing was an outlier on the cultural arena because as we saw in early modern times, chronicler/historians mention hurricanes and earthquakes in ways later literature and art foreclosed. Tapia in his *Biblioteca histórica* (1854) and *Mis memorias* (1928, posthumous), mentions 16 *temporales*. Brau in three books (1894, 1904, 1907) talks about 13 hurricanes. Coll y Toste mentions 43 hurricanes in *Boletín histórico* (published between 1914 and 1926). By 1950 climate historian Luis A. Salivia centers his monumental *Historia de los temporales de Puerto Rico* ("History of Puerto Rico's Hurricanes") on meteorological history. The most ecocritical history of cyclonic activity to that time, it combines science and history with stories of heroes and eyewitness testimonies of hurricanes. Salivia's tome details 94 *temporales* between 1508 and 1949 calling them "our" hurricanes, despite the fact that hurricanes are by their very nature "transregional transtemporal super-forces" (Hartman 2023, 123). For Salivia the environment is a crucial metanarrative to the story of the Nation.

On the other hand, Rafael Picó's *Geografía de Puerto Rico* ("Geography of Puerto Rico," 1954) lists hurricanes in seventh place behind what he considers factors of greater climate impact such as trade winds and easterly waves. Shockingly, he says: "fortunately there are few hurricanes that pass over Puerto Rico, but every year there are some passing close enough to affect our climate" (1954, 115). Our Appendix X conclusively proves otherwise.

Climate anxiety becomes a constant motif in Puerto Rican culture in the twentieth century. Both highbrow and popular culture tackle disaster: in poetry Palés's dancing storms, in art Tufiño's *La plena* monster, and in popular culture the *plena* "Temporal, temporal" ("Tempest, tempest"; Anonymous). This quintessential hurricane song references the devastating San Felipe II-Okeechobee of 1928. Its oft-quoted lyrics reiterate the theme of an anxious nation under climate threat: "¿Qué será de mi Borinquen cuando llegue el temporal?" ("What will become of Puerto Rico, when the tempest arrives?").

The cultural weight of the allegory of the garden-nation, however, persists at beginning of the twentieth century; for example Virgilio Dávila's poem "La Tierruca" ("The Homeland") (written in 1916, the year of Category 2 San Hipólito), offers memorable images of a quaint, small and verdant isle. He follows nineteenth-century religious tropes of a paradise lost: "paraíso terrenal, primavera, ¡Es Borinquen la mansión de todo bien!" ("lost earthly paradise, spring, Borinquen is the manor of all goodness!"; Quintana 1996, 287). Dávila's religiosity is deeply imbricated with the garden allegory of an unspoiled land of plenty and serenity.

Nostalgia was still a substantial driver for Dávila's vision of *la patria*, which becomes dearer when recollected from afar. In 1939, not much after San Ciprián, he writes from the point of view of a Boricua abroad, dreaming of an idealized landscape, in the aptly titled poem "Nostalgia": "En el invierno más crudo / ni un árbol se ve desnudo, / ni una vega sin verdor" ("In the deepest winter / not a tree is bare, / not a meadow greenless"; Quintana 1996, 249–250), because Dávila's enchanted land has Romantic flora and fauna of eternal springtime.

Myths of a placid landscape persist in poetry well into the twentieth century. Tomás Blanco personifies two "main" breezes in the island as: a nocturnal zephyr and a diurnal wafting wind.[3] Blanco gazes on the skin and breasts of both winds as female "lovers" in 1968, eight years after Category 4 hurricane Donna, yet ecoerasure still persists. Whither hurricane winds? There are also continuous uses of hurricanes as metaphors for poets' tempestuous feelings of love and yearning, which still appears in a poet as attuned to twenty-first-century concerns (LBGT + , migration, gender issues) as Yolanda Arroyo Pizarro, who in 2006 once again turns emotions into hurricanes: "Los ojos se me han vuelto mares huracanados" ("My eyes have turned into hurricaned seas").

However, traditional views of placidity cede ground to ecopoetical images of hurricanes, whose ferocious chaos bring climate anxiety, grief, and feelings of loss, as well as resilience. Leading twentieth-century poets display profuse symbolic uses of hurricanes as part of the Nation. Pro-independence activist poets use hurricanes as symbols of strength against colonialism, as well as its opposite, metaphors for colonialism itself. Francisco Matos Paoli, in a poem dedicated to independence icon Pedro Albizu Campos says, "[s]i queremos tocar el capullo de rosa, primero el huracán ha de abrir la esperanza" ("if we want to touch the rosebud, first the hurricane must unlock hope" n.d.). He dreams of independence, which like a hurricane, will renew the land.

Modernist Luis Lloréns Torres uses hurricane metaphors in his "Canción de las Antillas" ("Song of the Antilles," 1914) likening sovereignty to a personified hurricane. He also widens his environmental gaze to the entire Caribbean. "Canción" specifically personifies the Caribbean islands as "las hijas somos de un ciclón" ("we are the daughters of a cyclone," n.d.) casting hurricanes as agentic forces for Caribbean identity itself. His poem "El patito feo" ("The Ugly Duckling") links hurricanes explicitly to the independence ideal: "escucha la voz bravía / de tu independencia santa / cuando al cielo la levanta / el huracán del Caribe / que con sus rayos la escribe / y con sus truenos la canta" ("listen to the untamed voice / of your holy independence / when to the sky it is lifted by / the hurricane of the Caribbean / which with its lightning writes it / and with its thunder sings it" (1967, xliii). Additionally, Lloréns personifies hurricanes as the *patria* itself, because as Noel Luna says, "su poética del paisaje es una política de la 'nación'" ("his poetics of the landscape is a politics of the 'nation'"; 1967, 56–57).

Like many other poets Lloréns also problematically ties the landscape to feminine energies, personifying the islands as women with a comical personification of the landscape as a lady *patria*. One of his *décimas* (ten-line stanzas) presents a daintily feminine landscape: "Ya está el lucero del alba / encimita del Palmar, / cómo horquilla de cristal / en el moño de una palma" ("There is the star of dawn / on the tippy top of the palm grove, / like a crystal hairpin / in the bun of a palm tree"; Rivera de Álvarez 1983, 270).

Luis Palés Matos's images of dancing hurricanes have great joy, along with his swaying "disheveled palm women." Storms and the environment are center stage in his most famous poetry volume, *Tuntún de pasa y grifería* (translated as "Drumbeats of Kink and Blackness").[4] Begun in 1937, it grapples with the environment in gendered terms, for instance, the sea as a male laborer (with muscles and armpits) and the storm as a dancing mulatta. The book followed San Felipe II-Okeechobee (Category 5) in 1928, San Nicolás (Category 1) in 1931 and San Ciprián II (Category 2) in 1932. Alluding to recent disasters, seven of the volume's poems fire up storm imagery. Rafael Rodríguez Cruz says the pervasive imagery of hurricanes in Palés is second only to that of drums: "After the drum, the tropical hurricane is the most important lyrical figure in the Antillean poetry of Luis Palés Matos" (2022, 59). Drums and cyclones resonate vibrantly with sounds and musicality in *Tuntún de pasa y grifería*, fundamentally connecting the nation to its violent environment.

Palés's ecopoetics personify storms full of movement encompassing the senses: tornados with one leg, lunatic tropical hurricanes, and rotating hot and green cyclones. These destructive forces are striking characters leading to an apotheosis of gyrating women. Palés also personifies the climate as male, his metaphor of a hurricane as a demented inmate sprung from an insane asylum, blasting free of all restraints, is a timeless insight into the (at the time) unknown suddenness and powerful arbitrariness of meteorological phenomena. The entire passage is gyratory and dizzying:

> Sobre su pata única, vertiginosamente, / gira y gira el tornado mordiéndose la cola / en trance de San Vito hasta caer redondo. / Le sigue el huracán loco del trópico / recién fugado de su celda de islas, / rasgándose con uñas de ráfagas cortantes / las camisas de fuerza que le ponen las nubes; / y detrás, el ciclón caliente y verde, / y sus desmelenadas mujeres de palmeras. ("On its single leg, vertiginously, / the tornado twists and turns biting its tail / in a San Vitus trance until it falls to the ground. / It is followed by the crazed tropical hurricane / fresh out of his island cell, / ripping with cutting nails of wind blasts / the straitjackets that the clouds put on it; / and following, the hot and green cyclone, / and its disheveled palm tree women"; 2003, 58)

These wild images full of color and sound imagine characters whose most salient characteristic is extreme freedom with unrestrained power that cannot be contained.

Palés also feminizes the destructive power of hurricanes, linking cyclones to now-questionable stereotypes of Black women as fierce, impetuous, and untamable forces of nature. Mercedes López-Baralt, one of Puerto Rico's greatest literary critics, observes that hurricanes are deeply representative of his view of the Caribbean: "but the hurricane does not become mythical until it appears linked to the great Palesian contribution to the emblem of Antilleanness: the archetype of the dancing, sensual and libertarian mulatta" (2013). She calls this connection an archetype, however, its roots are racist stereotypes of bodacious Black women's bodies.

The storms can also bring the opposite to colonial stagnation. Palés ties cyclonic images of freedom to a pro-independence stance connected to mulattas/hurricanes: "Dale a la popa, mulata, / proyecta en la eternidad / ese tumbo de caderas / que es ráfaga de huracán,... / ¡para que rabie el Tío Sam!" ("Thump that rump, mulatta, / project to eternity / that hip beat / that is a gust of hurricane,... / to spite Uncle Sam!"; 2000, 118). Caribbean dancing is for Palés a defining cultural force. In his memorable *Tendambumba de la Quimbamba*, a Black woman dancing and carousing in the "encendida calle antillana" ("lit Antillean street"), rests on a vision of a Caribbean zone undergirded by hurricane energies.

Palés throughout underscores the toxicity of colonialism in Caribbean nations as the Antilles suffer natural and human-caused disasters. In other words, disaster nations. Hurricanes become the lynchpin of his Antillean metanarrative: "¡Antillas, mis Antillas! / Sobre el mar de Colón, aupadas todas, / sobre el Caribe mar, todas unidas, / soñando y padeciendo y forcejeando / contra pestes, ciclones y codicias" ("Antilles, my Antilles! / On the sea of Columbus, all uplifted, / on the Caribbean Sea, all united, / dreaming and suffering and struggling / against pests, cyclones and greed"; 2000, 82). Yet his vision is not hopeless; he opens *Tuntún* with "Preludio en Boricua" ("Prelude in Boricua") with a different, more positive and humorous take of the islands, depicted as characters swatting cyclones away: "Las islas barloventeras / pasan tremendas desazones, / espantándose los ciclones / con matamoscas de palmeras. / Como si no fuera gran cosa" ("The Windward Islands / have tremendous jitters, / Swatting away cyclones / with palm fly swatters. / As if it were no big deal"; 2000, 22).

Julia de Burgos's ecopoetics take the Puerto Rican landscape and its oceanic environment to a place far beyond mere aesthetic contemplation, enacting a passionate relationship with Nature and environmental justice. At times she uses storms in traditional ecophilic, even Romantic ways; for example, she herself is like a torrent, a "río crecido en plena tempestad" ("swollen river in the middle of the tempest"; 1997, 190).[5] Yet Burgos is an ecocritical poet *avant la lettre* in her regard for the agency of the non-human, forces that are characters. Her profound connection to the environment ushers in distinctive non-human elements to the landscape. Her ecopoetics consistently resonate with ecocritical material feminism; the basic premise of material feminism is: "the awareness that we inhabit a dimension crisscrossed by vibrant forces, that hybridize human and nonhuman matters, and finally the persuasion that matter and meaning constitute the fabric of our storied world" (Iovino and Oppermann 2014, 5). That is Julia de Burgos to a tee.

Burgos was a child when a tsunami and earthquakes struck in 1918. As a young woman, she lived through the disaster of San Liborio in 1926, and the monstrous catastrophe of San Felipe-Okeechobee in 1928. A further two hurricanes happened while she was still on the island, San Nicolás in 1931 and the much deadlier San Ciprián in 1932. Perhaps those disasters provided fodder for one of her most striking similes of how a woman is defeated by social norms, as she describes herself "como huracán cansado" ("like a tired hurricane"; 1997, 88), a comparison between the mildness required by society and her powerful interiority.

Her landscapes never stray from her anti-imperialist, pro-independence activism. Her early poem "Desde el Puente Martín Peña" ("From the Martín Peña Bridge," 1938) presents a broken earth, a physically and psychically shattered landscape within which the poet makes common cause with the downtrodden. The poem is a call to arms and unity among poor workers whose dwellings are "un ejército de casas," ("an army of houses"; 1997, 38), armed before blue sky and tranquil sea.

Chief among her environmental points of reference is the aquatic environment—one poem is simply titled "Mi poema de agua" ("My Water Poem"). Her most famous encounter with the aquatic environment, "Río Grande de Loíza," cemented her fame as an ecopoet of Nature. In it she melds her life story with her interlocutor, the Río Grande de Loíza, a vivid symbol of the environment that is her "íntimo secreto" ("intimate secret"; 1997, 8). Her body's material compounds inhabit the same symbolic space as the river. This river has agency as a character because her imaginary does not portray Nature as inert but as "endowed with many of the same qualities that were formally seen as exclusive to human beings: complex self–organization, reflexivity, consciousness, and the capacity to act spontaneously" (Bergthaller 2014, 37). Burgos's personified river is an explicitly sexual masculine lover, ground-breaking in its contrast with traditional poetry (for example Gautier Benítez), which feminized the landscape as a virginal island.

Her feminist underpinnings distance Burgos from the patriotic, apolitical nineteenth-century veneration of traditionally feminine beauties of the landscape. In an early poem, "Romance de la perla" ("Romance of the Pearl"), Burgos reinvents the clichéd image of the pearl, that gem from the sea, which critic Ivette López notes had been central to Gautier's "Edenic discourse" (2000, 107). Burgos melds social justice into her poem in reference to the poor neighborhood of La Perla. It's a double reversal since she also turns on its head the motto "La Perla del Caribe" promoted by commercial and governmental agencies.[6] This and her other political poems entangle the human and non-human, inverting the traditional gaze of idyllic spaces in order to denounce poverty and oppression.

Her personal experience of the nation's environment was heavily influenced by her exposure as a child and later as a teacher to the inland towns of the heartland, at the time largely untouched by modernity. One of her environmental justice poems, "Campo 1" ("Countryside 1"), represents a landscape scarred and engulfed by colonialism and injustice (1997, 344). She favors action over vision, a striking critique of traditional aesthetic contemplations of the serene Land; she continually proclaims that just gazing at the landscape is futile, humans must engage the natural world. A beautiful Nature is not enough if the fog of injustice engulfs the Land. A rallying cry ends "Campo 1", personifying the fight against injustice in the landscape: "¡hay que echarla del campo¡ /¡con los riscos, si faltan los brazos!" ("we have to eradicate it from the countryside! / with the cliffside, if arms are missing!"; 1997, 344).

She presents the poetic equivalent of Oller's bodegones in the poem "Campo 2" ("Countryside 2"), where she lists autochthonous Puerto Rican flora and fauna: tropical cherry trees, rose apple trees, butterflies, wind, puddles, hills, cliffs, boulders, thistles, a brook, lilies, *moriviví*, lemon tree, a guava tree (1997, 346). The profusion of *living* elements is in sharp contrast with the "trágico horizonte de piedra" (1997, 244) in her poem "Poema para mi muerte" ("Poem for My Death") when the poet is far from her homeland's environment, suffering in the cold New York cityscape where she will die.

Her spirited interaction with her environment runs throughout her oeuvre. Her *cri de coeur* "¡Oh mar, no esperes más!" ("Oh, sea, wait no longer!"; 1997, 202) addresses the sea, an interlocutor personified as a lover (as she does with the river in "Río Grande de Loíza"). "Yo misma fui mi ruta" ("I Was My Own Route") delves into the generative forces of earth as she becomes a natural force beyond history, transforming into a non-human seed sprouting from the earth into the cycle of nature.

We have explored how hurricanes can be transformed in poetry into the inner life of poets, the political life of the nation, and the reality of the Caribbean environment. These transformations prove how poetry manipulates real ecological phenomena.

Twentieth-century Puerto Rican prose also resoundingly abandons nineteenth-century myths of a benign Nature to deal with disasters, infrastructure, and the environment. For instance, starting in the 1940s José Luis González's stories feature descriptions of the countryside that are in stark contrast to old-style bucolic *paisajismo*, showcasing instead the harsh conditions brought on by U.S. sugarcane industries. He writes in the gap between hurricanes of 1932 to 1956 yet his is a hardscrabble countryside where the traditional idyll is replaced with decaying housing and poor, hungry people. The social realist landscape in "En la sombra" ("In the shadows") from 1943, includes mosquitoes, an environmental nuisance and health hazard that was never featured in traditional depictions of a benevolent Nature (1997, 70). Even nowadays mild ecophilic symbols of Puerto Rico's environment persist: *coquí*, parrot, palm, flamboyán, *ceiba*, and other picturesque items, whereas mosquito swarms are still mostly absent.

Many twentieth-century prose creators get stuck in the middle between the idyllic green *Isla* and denunciations of modernity's depredations of the landscape. For instance, nostalgia intertwines with the landscape in Antonio S. Pedreira's important, if dated, *Insularismo*, which sought to define Puerto Ricanness, in which he bemoans modernity for causing the loss of traditional greenery: "La tierra se nos va de las manos sin sentirla y los bosques han desaparecido a golpe de hacha, dejando llanuras y colinas huérfanas de árboles" ("We are losing the land without realizing it and woodlands have disappeared from axe strokes, leaving plains and hills orphaned of trees"; 1971, 47).

Human depredations of the traditional greenery alone don't justify his anxiety about a denuded landscape, what does explain it are the hurricanes of his era, which must have indeed left a stripped landscape. Born in 1899, Pedreira lived through San Hipólito in 1916, then earthquakes and a tsunami in 1918, San Liborio in 1926, the massive San Felipe in 1928, San Nicolás in 1931, and San Ciprián in 1932, two years before *Insularismo* appeared. The scars of disaster are obvious in his proclamation that Puerto Ricans are a sad people (1971, 41), in a land, "acosada de terremotos, de temporales y de impuestos" ("beset by earthquakes, storms and taxes"; 1971, 45).

One of Pedreira's subtitles is "Biología, geografía, alma" ("Biology, Geography, Soul"; 31) in which geographical determinism makes the main island's smallness define its people's character. His vision of a small nation, "operamos en diminutivo" ("we operate in diminutive"; 44), melds with his elitist belief that, "[n]uestra muchedumbre... es dócil y pacífica: se carateriza por la resignación" ("our masses... are docile and peaceful: characterized by resignation"; 1971, 42). For him the causes of Puerto Rico's melancholy, are, yes, the political tragedy of colonialism, but also the limiting Land.

The docility stereotype for Puerto Ricans has a long history; chroniclers already use it for Tainos. Bartolomé de las Casas describes Tainos in *Brevísima relación* as:

> los más simples, sin maldades ni dobleces, obedientísimas, fidelísimas a sus señores naturales y a los cristianos a quien sirven; más humildes, más pacientes, más pacíficas y quietas … que hay en el mundo. ("the simplest folk, without evil or duplicity, the most obedient, most faithful to their natural masters and to the Christians whom they serve; more humble, more patient, more peaceful and quiet … that there are in the world"; n.d.)

The chroniclers were, of course, selling the notion of an easily conquered people to the Crown in order to solidify their control over the indigenous population.

Pedreira reimagines this myth of docility, as does René Marqués who in 1963 (in)famously attributed the supposedly docile national character flaw to U.S. colonialism and what he erroneously called its matriarchy in "El puertorriqueño dócil" ("The Docile Puerto Rican"). Jorge Duany says this supposed docility consists of: resignation, fatalism, passivity, lack of initiative, conformity, repressed hostility, an inferiority complex, mistrust, and rejection of others ("Más allá de la docilidad"; 1991, 5). Other critics debunk Puerto Rican docility: Manuel Maldonado Denis in 1981 in "Un mito: La docilidad del puertorriqueño," Carmelo Rosario Natal in 1987 in *El puertorriqueño dócil: Historia, pasión y muerte de un mito.*

Puerto Rican docility is as fictional as the garden paradise allegory. So-called Puerto Rican docility is demonstrably false and repeatedly given the lie by: 1) the nation's fiery independentista movements;[7] and 2) its crime statistics, especially the current problem of gender violence. Arcadio Díaz-Quiñones further demystifies docility when he says the island has been a militarized zone since 1917, demolishing the view of a peaceable island by listing the popular appeal of many social benefits of widespread militarization: pensions, scholarships, educational opportunities, and mortgage loans (1993, 34). Pedreira was stuck in the middle between memorializing the violence of the environment yet still concluding that docility constituted part of the national character through a mix of politics, race, and bioregion, all of which he concluded lead to a cowering populace. He summarizes climate anxiety in Puerto Rico:

> por nuestra posición geográfica, vivimos en permanente angustia aguardando en los meses de verano las tormentas destructoras y en cualquier época los indeseables terremotos. Temblores y temporales nos sorprenden y vivimos en perpetuo acecho de cataclismos geográficos inevitables. Amenazados constantemente por la naturaleza ... hemos tenido que vivir expectantes y agónicos, acosados por desazones y derrotas. ... Esta actitud derrotista está vigente en nuestro general acatamiento y en arteria importante de nuestro pesimismo. ("because of our geographical position, we live in permanent anguish awaiting destructive storms in the summer months and at any time undesirable earthquakes. Tremors and storms surprise us and we live perpetually stalked by inevitable geographical cataclysms. Constantly threatened by nature ... we have had to live waiting and agonizing, beset by disappointments and defeats. ... This defeatist attitude prevails in our general submissiveness and is an important artery of our pessimism"; 1971, 45–46)

Pedreira connects Puerto Rican national identity with climate anxiety yet is also stuck midway between realistic acknowledgment of natural and human-caused disasters and traditional ecoerasure. He says disasters define the nation, nevertheless, he cannot escape the trap of a fictional landscape: "su nota predominante es la lírica: es un paisaje tierno, blando, muelle, cristalino" ("its predominant note is lyrical: it is a tender, soft, bouyant, crystalline landscape"; 1971, 37). This contradiction is evident when he states that the mild landscape resonates with its people: "Somos un pueblo ajeno a la violencia y cortésmente pacífico, como nuestro paisaje" ("We are a people oblivious to violence and courteously peaceful, like our landscape"; 1971, 41). Peaceful landscape? Whither San Ciprián?

Margo Arce de Vázquez is another writer caught between ecoerasure and ecorealism. Her essay "El paisaje de Puerto Rico" ("The landscape of Puerto Rico," 1939), toggles between the ecoerasure of a serene landscape and the history of hurricanes. She was born in 1904 and San Ciprián passed through the island in 1932 but more importantly, she was alive in 1928 when the island suffered San Felipe-Okeechobee (Category 5, by all accounts the mother of all storms until that time). Still, like Pedreira, she defines national identity through geography as determinant of the island's economy and its inhabitants. Her pessimistic view of the Puerto Rican character also leans into docility; the Puerto Rican is not "hombre de mar ni comerciante astuto" ("seafarer nor astute merchant"), furthermore, "como isleño, se deja ganar fácilmente por los aires de afuera" ("as an islander, he allows himself to be swayed by outside affectations"; Quintana 1996, 293). She also replicates sexism and Hispanophilic ideas embedded in descriptions of the land, citing its "gracia infantil" ("childlike grace"), and saying its greenery and hills look like toys. Gracefully, Arce extends the imagery of placid waters to encompass the land (Quintana 1996, 293).

Arce reaffirms an important theme: the resilience of a people who rebuild after disaster: "Cuando sobre la isla se desata la furia del huracán tropical, Puerto Rico sale de ese ritmo apacible, de esta dulzura de égloga. Pasado el temporal se vuelve a construir la casa arruinada, a cultivar el campo abrazado y a esperar con paciencia el nuevo desastre" ("When the fury of the tropical hurricane is unleashed on the island, Puerto Rico abandons that gentle rhythm, this sweetness of an eclogue. After the storm, one rebuilds the ruined house, one cultivates the razed field and one awaits the new disaster patiently"; Quintana 1996, 294). Hurricanes are proof of national resilience yet her 1950 ecoerasing poem "Isla" again feminizes the land, writing of its "maternal and young womb" amid a list of "authentic" vegetation (Quintana 1996, 118).

Arcadio Díaz Quiñones identifies "el arte de bregar" ("the art of coping") as a uniquely Puerto Rican social and intellectual trait yet also adds "el arte de borrar" ("the art of erasing"), giving as example, "[p]ara algunos, resulta tranquilizador borrar la experiencia brutal que han tenido muchos "ciudadanos" puertorriqueños debido a la discriminación racial y clasista en la metrópoli" ("for some, it is reassuring to erase the brutal experience that many Puerto Rican 'citizens' have had due to racial and class discrimination in the metropolis"; 2000, 13). Yet even though he writes in 2000 he ecoerases the region's violent climate and geology, as many Puerto Rican intellectuals had done before.

Many Puerto Rican creators partake of the global theme of the endless sea, a transcultural staple since Homer's wine-dark sea. For instance, Shakespeare's view of the sea as infinite in *Romeo and Juliet*: "Juliet: My bounty is as boundless as the sea, / my love as deep; the more I give to thee, / the more I have, for both are infinite" (II. ii. 133–135).[8] Emily Dickinson views the sea as utterly vast: "Exultation is the going / Of an inland soul to sea, — / Past the houses—past the headlands— / Into deep Eternity." Similarly, James Joyce's *Ulysses*'s "Ithaca" chapter devotes pages to the properties of water, at least two of which replicate images of a boundless sea: the watery world's "indisputable hegemony" and "its infallibility as paradigm and paragon." Belief in a timeless oceanic environment has, however, turned into "nature as an ahistorical phenomenon [that] makes it impossible to think about climate change" (Trexler and Johns-Putra 2011, 21).

Puerto Rican culture also has its share of the endless sea myth. Bobby Capó's 1952 song "Piel canela" ("Cinnamon Skin") proclaims the beloved cinnamon skin will persist even if the wide sea should lose its "immensity." Poet Evaristo Ribera Chevremont extols the enddless sea in "Mar Atlántico": "Sólo tú existes, mar. Sólo tú existes, / mar de inocencias y delicias únicas; / sólo tú, mar, existes. Mar de siempre... Y soy tú mismo, mar. Soy, en ti, eterno" ("Only you exist, sea. Only you exist, / sea of innocence and unique delights; / only you, sea, exist. Sea of forever... And I am you yourself, sea. I am, in you, eternal"; 1971, 37). He uses the myth in 1971, when the last hurricane had been Donna in 1960. The memory of the 100 deaths it caused was perhaps long gone.

As the twentieth-century progresses the myth of the endless seas becomes a pure fiction as the Capitalocene trashed the oceans, resulting in seascapes of, "contaminated fish consumption, plastic pollution, and radioactive waste in the oceans" (Opperman 2019, 446). Mega-fisheries recklessly trawl the oceans heightening the tragedy of the commons. Patricia Yaeger says the "endless sea" fiction contributes to a mindset that has led to the environmental crises of the Capitalocene. She understands why Emily Dickinson would dream of nature as Eternity but Yaeger says now the cost of this persistent myth is denying the: "technocean" in which "oceanic magnanimity turns toxic and has led to feckless oil drilling and overfishing" (2010, 530). Yaeger adds that potent poetic and artistic images "are part of a network or meshwork of ordinary ideas that echo throughout the oceanic commons with real economic effects" (2010, 531).

And yet the ecoerasing myth of the endless sea persists in Ana Lydia Vega's 2008 column "Mi país es el mar" ("My Country is the Sea"), in which Puerto Rican identity encompasses its aquatic surroundings (2008, 22). Her aquatic map re-charts the nation within the oceanic environment, giving the main island an oceanic vastness greater than the traditional land measure of 35 by 100 miles.

In the 1930s and 1940s storms were usually viewed through their effects on humans, especially the iconic *jíbaro* peasant. Miguel Meléndez Muñoz's 1936 story "Portalatín in Bankruptcy" mentions cyclonic impacts on coffee crops, but his ecophilic reliance on natural succession assures Nature's resilience: "la tierra quedó limpia" ("the earth was cleansed"; 1967, 101). He repeats this in the story "Los árboles" ("The Trees") describing how torrential rains are "revolutions" of Nature that defeat the *jíbaro* (1967, 21).

Enrique Laguerre's *La llamarada* ("The Blaze," 1935) appeared three years after Category 4 San Ciprián, and ecophobically treats Nature as an enemy. Hurricanes and the sugarcane industry are twin oppressors of the working class in this neo-realist novel that depicts terrible working conditions imposed by the sugar industry: back-breaking labor, little pay, corporate oppression, and fires. Hurricanes are a further tribulation, as if nature conspires with colonialism to crush people. *La llamarada*'s storm adds to the general anxiety with a ferocious metaphor of wild horses rampaging across the landscape (1994, 141). Laguerre's 1940 novel *Solar Montoya* shows the social and environmental consequences of hurricanes on the coffee industry and the *jíbaros* who depended on it. Laguerre alludes to three hurricanes: San Felipe in 1928, San Nicolás in 1931 and San Ciprián in 1932. He makes a strong connection between the landscape and the people whose love of a green landscape buckles under hurricane destruction: "Muchos montes se han venido al suelo de la primera tormenta. Tanto árbol destrozado... Los cerros se veían blancos con las hojas de los yagrumos caídos ("Many hills have collapsed to the ground from the first storm. So many trees destroyed... The hills looked white with fallen *yagrumo* leaves"; 1967, 169). The image of fallen white leaves like snow, is particularly striking in contrast to the pervasive cultural myth of a placid green land since it has political overtones as the "snow" of U.S. colonialism covers the land. *Solar Montoya*'s *jíbaros* are passionately connected to the environment. Doña Ana and her husband are attached to their beloved green landscape. She has an affinity to and identification with Nature and sees the beauty despite the thorns (1967, 110). Her ecophilia grants agency to the non-human; her husband, Don Alonso, shares the awareness of the interdependency of humans and Nature. Later Laguerre denounced the indiscriminate destruction of trees in sacrifice to the "devastador presentismo" ("devastating presentness") of progress (1974, 7A).

Migration from the wholesome countryside (the heart of the nation) to degraded cities, is a common topic in Puerto Rican literature in the mid-twentieth century.[9] *Solar Montoya* grapples with the impacts of hurricanes on increased migration when the protagonist Gonzalo leaves the beloved countryside for New York. His former boss writes him with ecophilic belief in a resilient nature and hope about recovering from hurricane losses, saying that a hurricane has defoliated the trees but that: "a pesar de las tormentas, se me sobran los ánimos. Cuando se quiere mucho la tierra, se hace de tripas corazón. Ya los palos tienen hojas nuevas y va desapareciendo aquel destrozo que me llenaba de tristeza" ("despite the storms, I'm in good spirits. When you love the land so much, you make it work. Already the trees have new leaves and the destruction that filled me with sadness is disappearing"; 1967, 149). We have seen that ecophilic hope since Abbad proclaimed in 1788 that the result of a hurricane is a "long serenity" and that destroyed plants will regenerate (430).

Miguel Meléndez Muñoz's *Cuentos del cedro* (1936) carries the same themes of disaster and reconstructive aftermath when the *jíbaro* protagonists battle "revolutions" of Nature while still believing that "la Naturaleza destruye para crear" ("Nature destroys in order to create";1967, 21). Concern for the *jíbaros* is still prevalent in 1973 in María Teresa Babín's story "Mañana es San Ciriaco" ("Tomorrow is San Ciriaco") which uses hurricanes as *costumbrista* markers of time. The hurricane season determines Puerto Rico's temporal life, how time is measured. The Spanish word *temporal* both relates to tempest and to time.[10] Historian Rudolph Adams Van Middledyk says hurricane activities helped islanders tell time, quoting an account in 1765 that: "they count time by changes in the Government, hurricanes, visits from bishops, arrivals of 'situados', etc." (1973, 112). Babín's story centers on the *jíbara* Jenara —born a time when birth certificates did not exist—and how her community measures time and memorializes the anniversary of San Ciriaco's passage.

Throughout the twentieth century the serene garden allegory would be discarded, as culture producers grappled with an ecorealist picture of the bioregion's hurricanes. Nevertheless, nostalgic feelings for a lost landscape appear up into the latter part of the century. Like Tapia in the nineteenth century, Edgardo Rodríguez Juliá laments a lost landscape: "[t]odo el paisaje de mi infancia ha desaparecido, y también el de mi adolescencia" ("all the landscape of my childhood has disappeared, and so has that of my adolescence"; 2005, 3). Infrastructure has overrun his remembered green landscape, which is now overdeveloped concrete. For example, an iconic, long-gone road from Aguas Buenas to Caguas, "una de las más hermosas del país, con sombra de pueblo a pueblo a causa de sus flamboyanes y jacarandas" ("one of the most beautiful in the country, with shade from town to town because of its poincinias and jacarandas"; 2005, 3). Lisa Paravisini-Gebert labels his attitude a "cantankerous nostalgia" that impacts national identity (2008, 8).

Two of Rodríguez Juliá's novels build images opposed to the placid garden. A tremendously prolific writer from the 1970s on, of his large *oeuvre* two texts wrestle directly with Puerto Rico's natural environment under colonial ecophobia. The first, *The Renunciation* (1974), was published during the gap in hurricane activity between 1960 and 1975; the second, *La noche oscura del Niño Avilés* ("The Dark Night of the Child Avilés"; 1984), not too long after the island was walloped by two hurricanes in 1979. Although neither novel features a hurricane, the latter has a tsunami and both take place in colonial settings rife with ecophobia. Both are set in a hallucinatory eighteenth century with racial struggles in phantasmagorical, baroque descriptions of the environment with human and non-human elements in turmoil.

It features an allegory of colonialism beneath an orderly garden paradise that is a deceptive and treacherous jumble. Rodríguez Juliá foregrounds the colonial Androcene objective of controlling the environment in a time and place inseparable from Catholic ideology. Exploding the garden as heterotopia Baltasar upends the hierarchy by making a double garden; an ecophilic "lovely and well-tended garden" hides another garden, an ecophobic labyrinth of horrors "beneath its beautiful vestments of flowers, singing birds, the light of the sun filtering through green leaves... lies the fruit of the discourse of evil" (1974, 55). Baltasar enacts his revenge on the ecoerasing colonial mindset, of which he says, "men spend all their lives in total indifference to the vegetable, mineral, and even human world," seeking to live only in, "ruined cities" (1974, 58). Monstrous ants and crabs are as much a part of the colony's environment as the selection of local fruits and vegetables that Bishop Larra gobbles down: soursops, medlars, *yautía*, sweet potatoes, yams (1974, 123–124) as he symbolically swallows the land.

La noche oscura del niño Avilés. Crónica de Nueva Venecia ("The Dark Night of the Child Avilés. Chronicle of New Venice")[11] has even more ecophobic and phantasmagorical imaginings of the environment. Nature is baroque excess; for instance, the tsunami that overtakes the fighting camp of bishop Trespalacios results not so much in human destruction, but in an ecocide for marine life (1984, 153). Rodríguez-Juliá's defamiliarizing ecocide is heightened with humorous notes as dying land-bound sharks start biting people. A similar incident happens when the rebel Obatal is in El Morro and a wave deposits octopuses and sharks on the lower walls, breaking down the human/non-human divide (1984, 162). Problematically, the novel identifies Black women with Nature, both considered equally uncontrollable. In a horror scene, El Renegado kills the Queen of Yyalode's pet manatee, which is described with female traits (1984, 178). As if deadly violence visited upon feminized fauna weren't enough, to prove his manly dominion over Nature he beats up *la reina*, the human woman, as well, so that women and the non-humans suffer equally.

Rosario Ferré, one of the most consequential writers of the last third of the twentieth century, portrays landscapes contaminated by the Androcene. Her phantasmagorical landscapes are shrouded in cement, that emblem of Puerto Rico's modernity which was a large factor in the fortune of her father, Luis A. Ferré, the first pro-statehood Governor. We could call him the twentieth-century Puerto Rican symbol of the Androcene. Ferré presents a political landscape in her ground-breaking 1976 *Papeles de Pandora*, published one year after Hurricane Eloise caused 44 deaths (Blake 2011). These stories stress capitalist overdevelopment in which Androcene oligarchs profit from the landscape. A male oligarch regards Puerto Rico's Nature as a strategic incentive for foreign investment saying:

> [h]aremos una consolidación de compañías y subiremos el precio diez centavos el quintal, solicitaremos ayuda federal veinte millones más, los traeremos a ver nuestras bellezas naturales y nos darán treinta millones más. ("we will consolidate companies and raise the price by ten cents per hundredweight, we will request federal aid for twenty million more, we will bring them to see our natural beauties and they will give us thirty million more"; 1976, 94)

Paradise equals profit. In *The House on the Lagoon* Ferré's exploited landscape compels Governor Winship who in 1937, the same year he ordered the Ponce Massacre (which is featured in the novel), proclaims himself "a lover" of the island's "fertility" (2014, 146). Winship wants to exploit the landscape for tourism, and in the tourist album he commissions: "there were angel-hair waterfalls, cotton-candy clouds, sugar-white beaches, cows pastoring up and down velvet-green hills—and not a single starving person marring the beauty of the landscape" (2014, 143).

Ferré's environment explicitly suffers from capitalist expansion so a dust-choked land and a polluted nature becomes "embalsamada en grasas de harinas grisáceas" ("embalmed in a greyish floury fat"; 1976, 109). *Papeles* prefigures the Saharan dust that now regularly covers the region, a phenomenon caused by the Capitalocene (Deopersad and Azevedo). For Ferré only a hurricane could clear the dust-canopied landscape as people wait for:

> la primera tormenta de rayos o huracán bendito que removiera aquel dosel de concreto armado que se elevaba sobre sus cabezas … la superficie de aquella tierra sembrada de cemento. ("the first lightning storm or blessed hurricane that could remove that canopy of reinforced concrete that towered over their heads. … The surface of that land sown with cement"; 1976, 95)

Her surrealist descriptions represent the denaturing forces of capitalist processes. In "El jardín de polvo" ("The Dust Garden") she upends the traditional allegory of Puerto Rico as a garden of eternal springtime since Ferré's asphyxiated, surrealist panorama is covered by "una capa de polvo que lo cubre dejando al descubierto la carne reluciente de venado" ("a layer of dust that covers it exposing the gleaming venison meat"; 1976, 102). Ferré transforms the olden garden myth into a bitter vision of a paradise spoiled by capitalism.

If Ferré's figurations highlight a surreal environment, her late twentieth-century peer, Magali García Ramis, is more reality-bound. Her environment features ecocritical material feminism, that is, texts in which humans (*la familia*) appear alongside the non-human as agents of life on Earth.

In "La familia de todos nosotros" ("The Family of All of Us") uncle Geño says the family has a moral disease, defeatism, not due to weakness, but from birth (1976, 80). Critic Myrna García-Calderón says: "For Geño as for Pedreira and later René Marqués, Puerto Ricans' main problem is their defeatism and docility" (1994, 59). When Aunt Iliana becomes ill the whole family steps in to help, each one bringing different remedies; Lydia and Geño use the most immediate and necessary medicine, they rub the sick woman's feet and hands with black, earthy soil (1976, 82). The rest of the family mock Lidia and uncle Geño when they say they cured Iliana with earth and stories, yet as the family's chroniclers Lidia and Geño connect with a Nature that belongs to "all of us" (1976, 85).

In her most famous work, *Felices días, tío Sergio* ("Happy Days, Uncle Sergio"), a hurricane is not metaphor, but a real-life threat for the family. The author says the novel refers to hurricane Santa Clara in 1956, because "it was the last –or one of the last– to have a saint's name, because afterward the Weather Bureau began to name them."[12] In 1953 NOAA's National Hurricane Center began giving secular names to Atlantic cyclones.[13] The hurricane in the novel symbolizes Puerto Rico's transition between overlords, with both a Catholic name, Santa Clara, and the name Betsy from the U.S.-based NOAA.[14] In that historical moment, when radar first began tracking hurricanes (Miner Solá 1995, 40), the family uses different kinds of knowledge. Straddling new and old worlds, *Felices días* vividly presents climate anxiety and modes of preparedness within the colonial context. The family, like the nation, has multiple forms of climate knowledge: folk wisdom, history, and tech-driven news. In the moment of dread when the storm is coming, crucially it is the family's women, starting with the grandmother, who connect to Nature to foretell the hurricane (1986, 43–44). When the radio finally announces the storm, the family prepares, closing doors, protecting windows, locking up animals, and stocking up on provisions. Lidia's mother, who remembers San Ciprián and San Ciriaco, pragmatically battles the storm preparing the family with lanterns and raingear, filling the bathtub with water, turning on the battery-powered radio, also listening to the news announcer Mr. McDowell.[15] The family also consult an old family map with the history of Caribbean hurricanes. At that moment of many transitions, their understanding of preparedness for hurricanes uses intuition, folk wisdom, history, and the scientific and technological know-how that will become routine. As disruptive as tío Sergio's sexuality, the threatening hurricane fractures the orderly world the family thinks is reality. Santa Clara/Betsy upends the placid bourgeois life in the time of Luis Muñoz Marín, yet for Lidia and the other children, the impending hurricane represents freedom beyond normalcy and conformity (1986, 47–48). The hurricane brings momentary freedom from social norms that suffocate the girl as well as the whole family. The storm shatters norms so much that Uncle Sergio openly helps the Tristanis, neighbors who are nationalists consequently "enemies" of the conservative family. As Larry La Fountain-Stokes says, "Nationalism in the novel is as monstrous as sexual 'aberration'" (2003, 54). Faced with the widespread dread of the hurricane, even the mother has a gesture of tolerance and compassion toward the neighbors as climate anxiety unites all Puerto Ricans.

García Ramis has special consideration for the non-human. The mother in *Felices días* appears non-human in the first paragraph when she is compared to a heron because of her one-legged stance. Lidia and her brother imitate her stance:

> balanceándonos en un solo pie, en una sola dirección, retábamos al mundo, al buen gusto y a la familia, mirando el sol caer, parados en fila como las garzas del río. ("balancing on one foot, in one direction, we challenged the world, good taste and family, watching the sun go down, standing in a row like river herons"; 1)

Here, when we first meet the core family, the narrative likens humans and non-humans. The mother's non-human posture captures one of García Ramis's main themes: nonconforming behaviors unacceptable to the *muñocista* bourgeoisie of the 1950s. The heron appears in several of García Ramis's texts and connects non-humans with humans. In the book *La R de mi padre* ("My Father's R") a solitary white heron takes flight to see San Juan from its point of view.

Her novel *Las horas del sur* ("The Hours in the South") also deals with the interrelationship between human and non-human, one of the pillars of ecocritical material feminism, which gives presence and agency to the non-human. The heron is again the emblematic non-human. The heron and her chick are mirror images of Monserrate and her son Andrés (2005,

49); human and non-human are both mothers integral to the seascape. When the rope that ties the boy on the beach comes loose and he floats away Monserrate calls for help, but there are no other humans around. The heron hears her and flies to the child, like an angel, and wakes him so he can swim back (2005, 50). *Las horas del sur* also has a hurricane; when Andrés is very young, the family takes shelter from a hurricane (that has, alas, no date in the novel). The boy is put in the shelter where Monserrate and the grandfather bring essential things, including the mother's treasures, threads and a crucifix, and the grandfather's books, tools, a compass and a card game. Seeking social order amid cyclonic chaos, they try to preserve from the hurricane (apart from the child who represents the future of their lineage) things that society dictates are gender appropriate by the social norms of Puerto Rican at the time.

A domesticated Nature reflects the conventional aesthetics and sense of control over the environment underneath middle-class living in Puerto Rico in the 1950s. The only birds at the beginning of *Felices días* are Mama Sara's chickens and pigeons, animals domesticated to serve humans. Later we see something worse, the rabbits that serve as experiments in the family laboratory. The family views the tropical rainforest El Yunque as tamed and tamable; they walk along lanes and eat next to placid creeks which, nevertheless, seem to come from "las entrañas de la tierra" ("the depths of the earth"; 1986, 14). However, when Lidia and her brother bury their dead cat, they dig in the earth and at that moment "iban apareciendo insectos extraños, caracoles largos y blancos, como alfileres finos y puntiagudos: piedrecitas pulidas, cementerios ignorados de criaturas minúsculas" ("strange insects started appearing, long white snails, like thin and pointed pins: polished pebbles, ignored cemeteries of tiny creatures"; 1986, 21). Connecting with the earth means strange insects and oddball creatures erupt, a non-human nature comparable to that "rare bird" Uncle Sergio, whose behaviors do not fit within *muñocista* heteronormativity.

García Ramis concludes that the *muñocista* era ravaged the traditional garden-like landscape planting cement everywhere in its colonized middle-class idea of progress:

> Eran años de cercenar montes de barro rojo para construir urbanizaciones, de abrir caminos de bitumul en cada monte verde, de florecer el cemento y los hoteles, de inaugurar represas y estaciones de electricidad y de esperar

> en el aeropuerto nuevo, que un día sería internacional, la llegada de americanos trajeados de gris tornasolado que mi tío Roberto iba a buscar a nombre de la compañía de foment. ("These were years of cutting down mountains of red mud to build housing developments, opening roads of tarmac in each green mountain, blooming cement and hotels, inaugurating dams and electricity stations and waiting at the new airport, which would one day be international, for the arrival of Americans dressed in iridescent gray who my uncle Roberto went to get for the development agency"; 1986, 2)

Green is identified with the autochthonous landscape and gray with notions of Americanized progress, like the Americans' clothing. Green holds a special place in the national imaginary beyond partisan politics. In a rare moment of appreciation of local nature, Aunt Ele—faithfully conservative member of a family that values everything European over all things Puerto Rican—proclaims that there is no better green: "'ni los verdes de Irlanda, ni los mismísimos verdes de Suiza, que es el país más impresionante del mundo' dice ella, 'se pueden comparar con los verdes de aquí'" ("'neither the greens of Ireland, nor the greens of Switzerland, which is the most impressive country in the world,' she says, 'can compare with the greens here'"; 1986, 24).

Landscapes represent starkly contrasting national identities in the novel. No matter how much Lidia recites Longfellow in El Yunque, she realizes she belongs in the tropical rainforest with no pines but instead prehistoric palm trees and giant ferns (1986, 139). The rainforest is deep and unknown and yet, "mío, más conocido y más parecido a mí" ("mine, more familiar and more like me"; 1986, 139). That green landscape is a vital part of the Puerto Rican identity she finally accepts. García Ramis criticizes, correctly, that the alternative national anthem *Verde Luz* ("Green Light") (which she admires) features an "isla doncella" ("maiden island") reaffirming the patriarchal prejudice of virginity as women's highest virtue: "that concept of maiden island is ideological machismo: virginity as a desirable value and the nation as a woman intervened and violated by another political group" (Negrón Muntaner and García Ramis 1993, 94). The song's nostalgic identification with the *patria*'s green light connects once again to a panorama of eternal springtime within climate, geology and women all regarded as serene.

García Ramis says that she and her contemporaries of the so-called Generation of 1970 study similar matters, although with different styles, adding that when they explore the middle-class ethos they do so: "with the same love with which natural scientists study sharks and remoras, which concern them, or the ecosystem that permits their survival" (1987, 111). Ecocritically, the role of writers is akin to that of scientists exploring the environment.

Ana Lydia Vega, one of the late twentieth-century triumvirate of feminist writers—along with Magali García Ramis and Rosario Ferré—critiques Androcene desire to control the environment. First, she sees a hurricane season as a socially unifying factor as well as marker of time, a weather season that governs life in the region. One could say that there are five seasons in the Caribbean: winter, spring, summer, autumn, and hurricane season. In "Encancaranublado," published one year after hurricane Hugo, Vega labels September a professional hurricane agitator that presages war (1990, 13), characterizing the environment as a battle between humans and that great character, the hurricane. Vega also recognizes that storms impact politics and create nationalist resentments. In 2008 Vega describes the anxiety when a storm approaches, re-establishing the temporal nature of hurricanes:

> septiembre ronda feroz por el Caribe como un dragón de 30 patas. Es la fecha fatal, el mes pico del invernazo, cuando la calentura máxima de las aguas envalentona a esos espíritus vengadores que resoplan ultimátums desde África. ("September roams ferociously across the Caribbean like a 30-legged dragon. It is the fatal time, the peak month of wintering, when the maximum heat of the waters emboldens those avenging spirits who breathe ultimatums from Africa"; 31)

Her personification of the hurricane as a dragon is reminiscent of Rafael Tufiño's monster in his mural *La Plena*. Vega presents the unifying power of climate anxiety, as an approaching hurricane becomes a joyful moment:

> Una alegría rebelde se va apoderando de los ánimos. Se brinda a nombre del huracán … se echan chistes a coro. El espíritu boricua se manifiesta en todo su esplendor pre-navideño. Por un instante confiscado al tiempo, el ciclón nos ha reconciliado. ("A rebellious joy takes hold of spirits. There are toasts to the hurricane, … jokes are made in chorus. The Puerto Rican spirit manifests itself in all its pre-Christmas splendor. For an instant confiscated from time, the cyclone has reconciled us"; 2008, 33–34)

One wonders, has that nervous elation in the face of a hurricane completely disappeared from Puerto Rican culture after Hurricane Maria?

"El baúl de Miss Florence" ("Miss Florence's Trunk") has a fictional hurricane as a plot device to expose the Androcene, mocking the sense of Man's control over Nature. The story comprises memoirs of a British governess in Puerto Rico in the mid-1800s and hews closely to her POV with its racist and class-conscious overtones, as well as her mawkish ideas about heterosexual love. Her journals cover 1856–1859 yet a hurricane strikes, even though between 1855 and 1867 there were no hurricanes. In contrast, Miss Florence sticks to another disaster faithfully when she recalls someone talking about the real-life cholera epidemic of 1855. So why does Vega invent a hurricane? To expose the Androcene; Mister Lind is a romance novel trope: the unattainable aristocrat who in the middle of the howling hurricane, when the slaves are penned in with the animals (which Miss Flo recounts without comment), heroically faces the storm to rescue an escaped horse. He returns drenched in rain, a picture of hunkiness, having saved the house (1991, 41). This is a sarcastic echo of Androcene heroics in which Man battles Storm, a masculinist trope as old as 1508 when Juan Ponce de León boasted of his Androcenic hurricane-battling skills. If Rosario Ferré speaks of a metaphorical hurricane, and in Magali García Ramis we see a material feminist view of Nature, for Ana Lydia Vega the hurricane unites people, marks time and intensifies plot while mocking Androcene masculinity. The most popular female triumvirate of their time had an ecocritical appreciation of their environment.

Art Embraces Hurricanes in the Twentieth Century

Depictions of a mild environment had been deeply tied to a coalescing national identity. Disguising the real environment, serene landscapes also cloak the poverty and disenfranchisement created by social inequalities and an extractive colonial economy. Scholar of Caribbean art Veerle Poupeye notes the link between depictions of a mild environment and money: "run-of-the-mill depictions of the natural environment tend often to be conventional in style and format, and usually present idealized, even stereotypical views of the Caribbean. Much of this work is mass produced in response to market demands, especially from tourism" (1998, 143).

Unless subsidized by the State, artists need to eat like any other worker and it pays to perpetuate a mythical garden paradise.

A change at a time of pro-independence fervor, artists in the 1950s cast hurricanes as major characters in life in a colonial system as twentieth-century Puerto Rican art progressively escapes the myth of an island paradise. As we saw with literature, some artists get caught in the middle between traditional patriotic (and commercial) views of a serene landscape and newer, ground-breaking ways to depict the violent side of the environment. Lorenzo Homar, a printmaker of towering importance in the twentieth century, shows in *Unicornio en la isla*[16] the persistence of ecoerasure. This 1965 print privileges Nature deemed authentic and autochthonous, as seen in patriotic representations of the landscape. In characteristic Homar fashion, the Tomás Blanco poem he quotes is as important as the artist's illustration of the landscape:

> Isla de la palmera y la guajana / en cinto de bullentes arecifes / y corola de soles. / Isla de amor y mar enamorado / los caballos azules con sus sueltas melenas / con desnuda piel de ascuas doradas / el torso de las dunas ("Island of the palm tree and the *guajana*[17] / In a belt of roiling reefs / and corolla of suns. / Island of love and besotted sea / The blue horses with their loose manes / with bare skin of golden embers / The torso of the dunes")

Like in Oller's landscapes, authentic plants like the palm tree are symbols of national identity. There is no trace of storms within that "besotted sea." The print's reclining landscape reminds the viewer of male gaze portrayals of women posing, especially with its reference to nude skin. The land is feminized in both Blanco's poem and Homar's print and identified with the mildest, most pliable environment; as fictional as the stereotypes of women. It is important to put this print in its ecocritical context: hurricane Donna in 1960 caused 100 deaths and much environmental damage. Yet the poem in 1964 and the print in 1965 represent a pristine, placid, bejeweled environment identified as "authentic":

> Isla de los coquis y los careyes / Con afrodisio cinturón de espuma / Y diadema de estrellas. / Isla de amor marino y mar embelesado. / Bajo los plenilunios: / Húmedas brisas, mágicas ensenadas, secretos matorrales... / Y el unicornio alzado / listo para la fuga alerta intense. ("Island of *coquis* and turtles / With an aphrodisiacal belt of foam / And a diadem of stars. / Island of marine love and enraptured sea. / Under the full

moons: / Humid breezes, magical inlets, secret shrubberies... / And the rebel unicorn / Ready to take flight, alert, intense")

What does this unicorn symbolize? It is as fictional as the eternally placid Caribbean climate that never really existed. The print's enraptured sea ecoerases meteorological reality by projecting sensual love unto the seascape. The words "diadem" and "aphrodisiacal" fully tip the poem, and Homar's loving homage toward a gentle environment undergirded by a softened, feminized Caribbean Nature. *Isla doncella* indeed.

Like most mid-twentieth-century creators Homar is caught in the middle, sometimes viewing Nature as benign, for example in the beautiful delicacy of his engraving *Picaflor* (Hummingbird").[18] Nevertheless, his print from 1971 quotes José de Diego's 1904 "En la brecha" ("Into the Breach") and has no figurative elements besides its font, consisting entirely of a quote from the poem with one of the most famous independentista uses of "storm" in Puerto Rican letters. Likewise, Homar's political cartoon print *Plenas: Tintorera de mar* ("*Plena* Song: Sea Shark" 1954)[19] shows nature's revenge on colonialist society as a shark gulps down a well-dressed lawyer-like figure. It is possible also to interpret his print *La tormenta* ("The tempest" N.d.)[20] as a political metaphor privileging humans since the subject is the man in the foreground while the storm is limited to dark skies in the background yet the dread of the hurricane, metaphorical or otherwise, comes across (Fig. 4.1).

Fig. 4.1 Rafael Tufiño Figueroa. *La Plena*. Casein/masonite, 1952–1954. Instituto de Cultura Puertorriqueña Collection

Rafael Tufiño Figueroa's monumental *La plena* ("The Plena Song" 1952–1954) is an artwork as ambitious in depicting the Nation as Francisco Oller's *El velorio*. Oller pictures a placid, green, domesticated landscape seen through doors and windows, even as the emblems of colonial society are indoors. Tufiño, on the other hand, personifies the hurricane as a green monster at the very center of the Nation, which is depicted through eleven other *plenas*, many representing violence against women.[21] The monster represents the *plena Temporal, temporal*, that most famous song about disaster (San Felipe II in 1928); at least until Bad Bunny's *Una velita* (Maria in 2017). The accompanying book of engravings, *12 grabados: La Plena* ("Eleven Prints: La Plena"), pairs Tufiño and Lorenzo Homar and, as Joseph R. Hartman says, portrays the hurricane as a "god-like figure [that] sweeps away humble masses of men and women in one hand while reaching for clustered wooden homes with the other" (2023, 124). Tufiño's monster also references Taino art, offering a deep connection to the hurricane "a shifting and powerful time–space with deep cultural roots" (Hartman 2023, 126). The roots go to the dawn of Puerto Rico's disaster culture in Taino art.

The mid-twentieth-century explosion of art would depict ecoreality as Puerto Rico's violent environment appears in numerous artworks that exploit the shock and awe of hurricanes while highlighting the socio-political underpinnings of disaster. The Puerto Rican government's División de Educación de la Comunidad ("Division of Community Education"), or DIVEDCO, was established in 1949 in part to promote hurricane preparedness through graphic arts and films.[22] The Instituto de Cultura Puertorriqueña's online exhibit DIVEDCO[23] singles out two hurricanes, San Ciriaco and San Ciprián, as weather catastrophes that impacted economy, art, and life. Poupeye states that in the first half of the twentieth-century "Puerto Rican art had been practically dormant" (1998, 113).

Hurricane art specifically denounces capitalist exploitation of disaster relief and reconstruction funds. Carlos Raquel Rivera's 1955 print *Huracán del norte* ("Hurricane of the North" ICP COP.1991.0699 (a-b)) shows naked people and objects flying mid-air in the background and in the middle ground the skull-shaped head of an anthropomorphic hurricane. The foreground has a wooden country family home, symbolic of all that the hurricane can destroy. A bag with coins spills within the hurricane to remind the viewer not just of the cost of hurricane damages but also of Androcene predatory disaster capitalism, as the figure of a businessman clings to the hurricane. Joseph R. Hartman calls this print "political satire" (2023, 137), and Poupeye says it offers a metaphor for "the destructive influence of North America, American materialism in Puerto Rican Society, one of the first openly anti-American statements in Puerto Rican art" (1998, 114). Hartman adds "U.S. imperial officials, Caribbean despots, and local politicians alike used architecture and visual propaganda to restore, reshape, and control cities destroyed by the worst hurricanes" (2023, 136). *Huracán del norte* is from 1955, long before Naomi Klein's 2008 *The Shock Doctrine* summarized predatory disaster capitalism. Her widely-used concept has been challenged though, for instance Danielle Zoe Rivera questions its use for a colonial state like Puerto Rico (2022, 128–129) (Fig. 4.2).

Hurricanes cause migration and climate refugees in José M. Figueroa's 1958 poster *Huracán* (ICP COP.1992.3133), which shows an anthropomorphic hurricane hovering in the sky, blowing over a battered climate refugee family that flees desperately, seeking shelter from the personified

Fig. 4.2 José M. Figueroa. *Huracán*. Silkscreen, 1958. Instituto de Cultura Puertorriqueña Collection

hurricane, a character as important in the composition as the humans. In the far background is the evidently fragile house they are trying to reach. Figueroa's poster combined with DIVEDCO's 1958 documentary *Huracán* and its accompanying book, *¿Qué sabemos del huracán?* ("What Do We Know about the Hurricane?"),[24] taught people living in vulnerable areas about precautionary planning needed before, during, and after a hurricane. The documentary contrasts favored new technologies against popular wisdom which had been used to forecast storms since early times, citing examples like interpretations of the movements of birds and the fullness of avocado crops[25] (Fig. 4.3).

1965 Eduardo Vera Cortés poster *Huracán* (ICP COP.1990.0477 (a-f)) builds on Raquel Rivera's and Figueroa's posters. Dominating the sky, this hurricane is more stylized and also scientifically correct, with counterclockwise motion.[26] Vera Cortés repeats the dramatic image of a family of climate refugees fleeing the storm that dominates the landscape. The family, that traditional heteronormative emotional center of the Nation, tries to escape the menacing storm about to hit "El Fanguito," a poor community built on the shores of a lagoon. A solitary concrete structure in the background reminds the audience of the island's rapid modernization in the 1950s and 1960s and also the government's campaign to deter the construction of unsound homes, which were, and are, highly susceptible to hurricane damage.

The ecoerasure of hurricanes and disaster we saw in much of Puerto Rican culture disappears almost entirely after 1950. The stories told, the images formed, the symbols deployed, are tied to the double disasters of hurricanes and ever-present colonialism. The destructive force of hurricanes became increasingly tied to the forces of colonialism and capitalism. Governmental action was not only important for DIVEDCO, preparedness and resilience has been tied to Puerto Rico's territorial status vis-à-vis the climate authorities of the United States ever since.

Hurricane works by Raquel Rivera, Figueroa and Vera Cortés had didactic intent. Rafael Tufiño's *La plena* personifies the hurricane as a main character of the Nation. His *Huracán* ("Hurricane"),[27] from 1971, would return to the hurricane as monster. It references Puerto Rico's deep culture by hearkening also to Taino mythology, depicting a male figure floating gently amid Taino symbols, its *cemí* pottery connects to disaster art in ancestral times.

Another notable figuration of a hurricane is Domingo García's 1994 sculpture *Juracán dios del viento (obra primera)* ("Juracán God of the

Fig. 4.3 Eduardo Vera Cortés. *Huracán*. Silkscreen, 1965. Instituto de Cultura Puertorriqueña Collection

Wind (First Work)"; ICP CPE.2021.1140) which again shows the lasting cultural connection with Taino heritage as well as representing the hurricane as a living organism with wings.

Puerto Rico's infrastructures offer proof of the fraying social contract under colonial mismanagement, which has been mostly ineffective in coping with disasters. Infrastructures have their conceptual roots in enlightenment ideas of progress (Larkin 2003, 328). Hurricane Maria in

2017 demolished such ideas of progress in Puerto Rico. Critical infrastructure that rests on a fossil fuel capitalist economy, particularly the dysfunctional electric grid, was incapable of recovering in 2022 from Category 1 hurricane Fiona. Manuel Avilés-Santiago traces this technological underdevelopment back to Spanish times saying that, "colonizing powers assert their monopolistic access to cutting-edge technology at the expense of the colonized (delayed technology transfer, for instance, or unilateral decision-making on the part of the colonial government concerning implementation)" (2015, 11). The persistence of the theme of colonial mismanagement of disaster goes back to 1679's anonymous letter comparing colonial regimes with hurricanes (see Chapter 1).

Infrastructure first appears in Puerto Rican art with José Campeche's 1792 painting *El Gobernador Don Miguel Antonio de Ustáriz.* Its influence is still apparent in 2003 when Rafael Trelles reinterprets *Ustáriz* in his painting *Todo se hace por ti*, which reproduces Campeche's work but changes Ustáriz's race as the text on the right lists vital State infrastructure: roads, schools, and hospitals. Oller's *hacienda* sugar mill painting series are important for the key place they give sugar mills as infrastructure, connecting his sublime skies (always free from hurricanes) with the social and economic life of the nation. The fact that the greatest national painter of the nineteenth century devoted six paintings to sugar mills is a sign that he viewed landscape and infrastructure as intertwined.

Roads, with their cement and sound pollution, are the ambient environment of everyday Puerto Rican life. Transportation infrastructure—the life flow of this petroculture nation—relies on roads, the most glaring part of Puerto Rican infrastructure. Of all the ills that can be laid at the feet of Luis Muñoz Marín and the Popular Democratic Party (PPD, for its initials in Spanish) one often forgotten is their destruction of the existing railway system. Historian Elba M. Tirado de Delucca documents how when the U.S. arrived the main island already had 370 km of railways which had been started in 1882 and by 1898 crisscrossed the main island, eventually encircling it (1997, 171). The railway benefitted the sugar industry yet also impacted the lives of people for the better, allowing public transportation and mobility, as well as better postal delivery; that all ended in the PDP's decades-long control. Tirado blames the loss on the need for faster transportation: "new needs arise that make the use of faster moving vehicles essential. Puerto Rico gets filled with private and public cars, making it necessary to expand streets and highways" (1997, 174). The influence of the U.S. auto industry was perhaps more to blame.

Essayist Arcadio Díaz-Quiñones says that the forms of political (ir)rationality behind this shift in transportation infrastructure during *muñocista* modernity had dire consequences (1993, 174). The island's landscape, infrastructure, and people still suffer from this massive historical blunder—as the costly, inefficient Tren Urbano demonstrates. How different would the Puerto Rican landscape, and people's lives, be with a working railroad system that encircled the island? Since the railway system was destroyed by the ELA under the influence of the U.S. auto makers, the island developed a road system that is a vital piece of infrastructure, the lifeblood of commerce and of how ordinary people get around but one that hurricanes and earthquakes regularly devastate. Therefore, a constant theme in twentieth-century Puerto Rican culture is overdevelopment, seen from René Marqués's dire depiction of life for the poor in the city to what Ana Lydia Vega terms "ecocide" in a column lamenting the pollution of air, sea, and land, and the extermination of flora and fauna (2011, 37).

Infrastructure is a main character in Luis Rafael Sánchez seminal *La guaracha del Macho Camacho* ("Macho Camacho's Beat," 1976), an allegory of the island as a traffic jam. Famously, the entirety of the novel takes place during the monumental traffic jam. Sánchez wrote when road dysfunction was already the most evident feature of the landscape. In *La guaracha* the wealthy characters drive expensive cars while the poor travel in the chronically malfunctioning and polluting bus system. Another vital piece of infrastructure that Sánchez also shows in crisis is housing, especially for the poorer characters who suffer most from maladapted infrastructure.

Other infrastructure dysfunctions, such as chronic water and energy insecurity, are aggravated by the interconnected inefficiencies of local and U.S. bureaucracies, worsened when disaster strikes. Puerto Rico offers many examples of what Patricia Yaeger calls maladapted infrastructure. She details the need for alternatives to combat crisis factors:

> (1) the fact of overurbanization, (2) the predicament of decaying or absent infrastructures, (3) the unevenness of shelter (which, along with food, energy, health care, and water make up the mythos and ethos of the nurturing city), and (4) the importance of inventing counterpublics, or communal alternatives to the official, bureaucratized polis. (2007, 13).

All these are common environmental and social challenges of life in Puerto Rico.

What the transformation of the landscape meant for stories was an increased awareness of the effects of infrastructure on both environment and people. As stated earlier, José Luis González's 1950 story "En el fondo del caño hay un negrito" ("At the Bottom of the Canal There Is a Little Black Boy"), proves that the *caño* suffers from colonial and capitalist "slow violence," an attritional, nearly invisible product of "violence that is neither spectacular nor instantaneous, but rather incremental, and accretive, its calamitous repercussions playing out across a range of temporal scales" (Nixon 2011, 2). The plague of Puerto Rican roads during modernity was a form of slow violence incrementally causing the *caño*'s deterioration. Rob Nixon says we need "stories dramatic enough to rouse public sentiment and warrant political intervention" (2011, 3). Stories like "En el fondo del caño hay un negrito." Happily, the rehabilitation in the 2010s of the *caño* is a prime example of community action for environmental justice.

There are many Puerto Rican writers and artists who focus on cityscapes and their inadequate infrastructure. Eduardo Lalo is a city writer and photographer who, as often happens in the late twentieth century, turns his back on the landscape. His work, and that of younger writers like Janette Becerra, elides the traditional landscape. They present Puerto Rico in the late twentieth and early twenty-first century as a nation where the city replaces the *campo* in an altered national landscape. When Lalo leaves the capital, he sees a natural environment through nostalgia. In *Donde* ("Where") he follows Tapia and Rodríguez-Juliá in mourning an old road which is now: "moteles, garajes de mecánica, talleres, bares, taquerías, que forman la tierra de nadie de la expansion de la capital" ("motels, garages, repair shops, bars, taquerías, which form the no man's land of the capital's expansion"; 2005, 86). In *La inutilidad* ("Uselessness") Lalo observes the dependency on cars stating that he and his friends rarely left the capital because they lacked reliable cars. Lalo is a Europeanist writer who in many ways continues the line of nineteenth-century writers like Tapia whose claim of cosmopolitanism fits Lalo, who writes as much about Paris as about San Juan in *La inutilidad*. In fact, he states that in San Juan, "me convertí, sin proponérmelo, en un elemento del paisaje" ("I became, without trying, an element of the landscape"; 2004, 178), meaning he faded into the background, hence communicating an insouciant dismissal of the traditional landscape in favor of complex cityscapes. His poem "San Juan by Night" in *La isla silente* has a *positive* look at city life right after a hurricane when, due to a power failure, the writer (at his most luddite or misanthropic) celebrates that no TVs are on and rejoices about "esta noche recuperada" ("this recovered night") in candlelight, even though he mentions flooded houses and millions in damages (2002, 155). Published in 2002, this take may be somewhat understandable since there were damages but, most importantly, no deaths from Hurricane Lenny in 1999 or Hurricane Debby in 2000.

A writer may ignore the bioregion altogether, setting her tales in foreign environments; Janette Becerra does so in her 2011 story, "Afición por los terrarios" ("Fondness for Terrariums"). Like many Latin American writers of the early twenty-first century who place their works outside their countries, Becerra's setting is not the Caribbean. The protagonist crafts terrariums, miniature alternative environments. Bored with his creations, he plays god by inducing disasters; first introducing a virus to see how the ecosystem copes with "el infortunio de las plagas y la calamidad" ("the misfortune of plagues and calamity"; 2011, 77). This cataclysmic disposition results in frightful animal deaths but ultimately brings about a hopeful renewal, as the mice grow resistant to viruses. Next the disaster builder makes a tempest with floods which cause great destruction; however, nature renews and the prodigious terrarium bursts again with strange new life forms, a trust in ecosystems' ability to self-regulate through natural succession.

The environmentally destructive force that is humankind, however, having caused the Capitalocene, will now suffer more extreme weather events. Lijing Cheng et al. summarize the problem of unprecedented oceanic warming: "Increased evaporation leads, in turn, to moistening of the atmosphere, released as latent heat during precipitation, invigorating the hydrological cycle" (2022, 776). That invigoration will most probably lead to "increased storminess" (2022, 782) and supercharged storms. Even more alarmingly, "ocean warming threatens the fundamental conditions that make the Earth conducive to sustaining life" (Cheng 2022, 777).

The Blue Humanities, necessary for studying Caribbean cultures, turn scholarly attention to "human engagements with liquid environments" (Blackmore 2020, 2). Scholars of aquatic environments (DeLoughrey, Yaeger, Blackstone, and others) stress the need "to put the oceans at the forefront of our attention" (Mentz 2019, 384) in literature and the arts to reflect back to readers and viewers the destructive effects of humans on our aquatic environments.

Traditionally, as we have seen, Puerto Rican literature and art romanticize the Caribbean Sea and stay away from its cyclones. Yet Edgardo Sanabria Santalíz turns this aquatic attitude on its head in the story "Después del huracán" ("After the Hurricane," 1983) when a hurricane leaves behind an inverted world where the ocean invades a home (Barradas 1983, 156–169).

Sanabria's phantasmagorical vision of human habitats altered by a hurricane inverts the normal human vs non-human world. The hurricane brings the surreal image of fish in place of leaves "ornamentando el paisaje" ("adorning the landscape"; Barradas 1983, 170) and echoes the destruction of the two hurricanes in a row that had struck four years before the story's publication, one of them, Frederic, was a Category 4 disaster. The intervention of the non-human fits into Elizabeth DeLoughrey and Tatiana Flores's theory of the ocean "not as blank space or aqua nullius for human agents to cross but as a viscous, ontological, and deeply material place, a dynamic force, and an unfathomable more-than-human world" (2017, 163).

Sanabria's is also a tale of Androcene abuse as the human traps, defiles, and exploits the siren Melusina, turning her into a paying attraction. His actions mirror what humans have been doing to the oceans in the Capitalocene. Patricia Yaeger explains the destruction of the oceans in our time, saying that "late-capitalist seas are becoming more techno than ocean. The Pacific gyres are dense amalgams of thrown-away plastic that form an aggregate twice the size of Texas" (2010, 527). Sanabria's story ends in nature's revenge when Melusina drags the protagonist underwater, as the real-life Capitalocene's rising oceans will drown so many inhabited lands.

Despite all we know, even at the end of the twentieth century it was still possible to ecoerase the environmental disasters of the Capitalocene. Storms would—will—become more erratic and powerful and plentiful yet in 1996 Mayra Montero humorously speculates on running out of names for these "vientos incordios" ("nuisance winds"; 1996, 126) in her story "Ciclones con nombres tiernos" ("Cyclones with Tender Names"). It doesn't seem so funny in the twenty-first century when we've twice run out of hurricane names (Whang 2020). Montero's light-hearted joke about "nuisance winds" tracks with her obliviousness about hurricane technology. She mentions two forecast technologies, barometric pressure and flights over hurricanes, but shouldn't Montero—and everyone who lives in the Hurricane Alley that is the Caribbean—know more about hurricane tech? Unless one is into hurricane data, there really is no need to use differential equations to study storms, but shouldn't a firmer knowledge of the scientific and technological complexity of hurricane forecasting inform the writing of a Caribbean writer like Montero? Yes, the barometer was an important advance that warned island nations of hurricanes, but that was in 1837! Montero mentions planes, yes, but hurricane hunter planes have been around since 1944. By 1996 Puerto Rico's preparedness offices counted on well-publicized computer advances to perform statistical and dynamical modeling for hurricane forecasting (1950s), radar (1956), hurricane-tracking satellites (1961), unmanned aerial vehicles (UAVs) and drones (1960s), planes with on-board radars and dropsondes (1970s). Ecocritical theories linking science and the humanities were in their infancy, but Montero using "planes" alone ecoerases nearly all hurricane know-how, which is grating after 1995's Hurricane Marilyn, which was a Category 3 phenomenon. Even at the end of the twentieth century Montero reproduces a traditional version of ecoerasure.

Notes

1. As opposed to commercial art and publicity-related material.
2. In March 2019 Utah Republican Senator Mike Lee introduced a bill to repeal the Jones Act, but failed to gain support in the Senate since it was met with fierce opposition by shipping industry groups such as the Florida Maritime partnership and the American Maritime Partnership. A tirade by industry group MEBA Union was titled, "Senator's Bill would Surrender U.S. Jobs to Foreign Shipping Interests."
3. "Ditirambo decorativo de las brisas" ("Decorative Dithyramb of Breezes" 53–56).
4. Apart from this widely-used translation for the title, all other translations are our own.
5. Our own translations.
6. Although Cuba would dispute the title.
7. Marqués attributes this political movement (which contradicts his whole premise) to a "suicidal" impulse (1977, 163).

8. On the other hand, Shakespeare uses storms strategically and in highly symbolic manner in *The Tempest*, of course, but also in *King Lear*.
9. René Marqués, for instance, emphatically embraces that migration trajectory in *La carreta* (1963).
10. It comes from Latin *temporalis*, composed of the suffix -*alis* (-al = relative to) on the word *tempus* (time, duration, period).
11. Our translations for this novel.
12. Direct message from the author, Aug. 2023.
13. An international committee of the United Nations World Meteorological Organization maintains and updates the system.
14. In connection to this feminization of hurricanes, traditional Catholic names used both male and female saints whereas from roughly 1953 to 1979 U.S. climate agencies only used women's names. Speculation was that this followed old maritime customs. It took protests by feminists such as Roxcy Bolton to persuade the agencies to start using male names again in 1979.
15. Noting the importance of television for hurricane preparedness, historian Fernando Picó proclaimed McDowell the living symbol of the weather for an entire generation (2008, 91).
16. https://web.archive.org/web/20250212182842/https://www.mapr.org/es/arte/obra/unicornio-en-la-isla.
17. The white flowering sprig of the sugar cane.
18. https://web.archive.org/web/20250212183433/https://dpul.princeton.edu/ga_treasures/catalog/9z902z98f.
19. https://www.moma.org/collection/works/31937.
20. https://web.archive.org/web/20250212184434/https://dpul.princeton.edu/ga_treasures/catalog/g445cd28v.
21. The other 11 *plenas* are: "Santa María," "Cortaron a Elena," "Tintorera del mar," "Isidora," "Ay, Lola, Lola," "Qué bonita es," "El Obispo," "Los muchachos de Cataño," "Ya llegó," "La guagua," "¡Fuego, fuego, fuego."
22. Veerle Poupeye highlights printmaking as an important tool for mass distribution of art, aided by the Center of Puerto Rican art (CAP) established in 1950 by Homar and Tufiño among others (114). CAP closed in 1952 but by then printmaking was firmly established as a popular artform allowing DIVEDCO to reproduce art quickly and at lower cost.

23. https://web.archive.org/web/20250212184835/https://artsandculture.google.com/story/hwXBjXRjMgt9NQ.
24. https://web.archive.org/web/20220705164547/https://expodivedco.sagrado.edu/que-sabemos-del-huracan-1958/.
25. This well-established myth that an abundant avocado crop portended a hurricane is reported by Teodoro Vidal, (2010, 27–30).
26. Hurricanes, or tropical cyclones, spin in a counterclockwise motion.
27. https://web.archive.org/web/20220929193937/http://www.prdream.com/galeria/tufino/huracan.html.

CHAPTER 5

Post-Maria Representations of Disaster, Testimony, Decolonizing Resilience, and Reconstruction

"El político es el peor huracán que tenemos"
("The politician is the worst hurricane we have")
Clemente González (taxi driver, 2023)

Carlos Ruiz Valarino's print *Sin título* ("Untitled," 2008) is a magnificent precursor to recent depictions of disaster in Puerto Rican art. This photographic composition prefigures two major challenges for Puerto Rico: climate change and migration. Nine years before hurricane Maria, Ruiz Valarino anticipates present crises with his image of people crowding a sinking ship's crow's nest (Puerto Rico in the rising waters of climate change) while looking at an airplane (migration) (Fig. 5.1).

Like a super-charged storm, Puerto Rican culture nowadays continues using the metanarrative of disaster. Climate change and energy justice suffuse contemporary Puerto Rican culture so that post-Maria artivists continuously call for climate justice. Ecoerasure is almost entirely untenable since hurricane Maria walloped the island in 2017,[1] reminding Puerto Ricans of the impact of catastrophes on every being in the landscape. As we saw in the last chapter, hurricanes reclaim their crucial role as key characters in the story of the Nation during the twentieth century; Maria's aftermath compounded their protagonism.

Early twenty-first-century culture still had a measure of ecoerasure. An anthology from 2011 called *En el ojo del huracán* ("In the Eye of the

M. Acosta Cruz, *An Ecocritical Study of Puerto Rican Culture*,
https://doi.org/10.1007/978-3-032-04044-2_5

Fig. 5.1 Carlos Ruiz Valarino. *Sin título*, photographic series *Proyecto Paradiso*, 2008. By permission of the artist

Hurricane") deals with "hurricane" entirely as a metaphor. It is a volume about the "I" of the hurricane, not about weather. The anthologists, Mayra Santos-Febres and Daniel Darío Carrero, explain how the "eye of the hurricane" metaphor comes from meteorology: "la pared del ojo del huracán es una zona donde se encuentran dos fuerzas opuestas entre sí: la fuerza del aire, que se mueve hacia el centro, y la fuerza centrífuga, que es hacia fuera" ("the wall of the eye of the hurricane is a zone where two opposing forces meet: the force of air, which moves towards the center, and the centrifugal force, which moves towards the outer edge"; 2011, 13). The anthology's topics range far and wide: Spanish dictatorship, music and nostalgia, Daleks, pop bands, Jesus Christ in Cuba, and "Mundo cruel" ("Cruel World") Luis Negrón's famous humorous story. The stories examine the full panoply of modern life except that is, real-life hurricanes. The environment's secondary role may be because the stories were gathered during the hurricane hiatus between 2000 and 2011.

Soon, however, the environment would not just be background; the same year the anthology appeared hurricane Irene caused $5000 million in damages. Consequently, community resilience rose as a literary topic. In *El huracán y la subjetividad antillana* Rafael Rodríguez Cruz reminisces (in 2022) about how in the 1960's a storm unified his black community in the town of Guayama. He highlights two characters, his grandmother and a hurricane that passed south of the island. This was probably hurricane San Lorenzo, or Donna, which caused 107 deaths in P.R. and St. Thomas. He says townspeople accepted popular beliefs such as the avocado crop forecasting the storm season. Sharing the religious beliefs from early modern times, his grandmother stated that San Felipe was devil's work, and retold the auditory and visual clues which she said revealed the presence of Satan during that storm.

Looking back on 1867's San Narciso, Adrián Bermúdez Pérez says (in 2022) that hurricanes deserve an important place in Puerto Rico's history. We should link that protagonism to other national disasters, particularly the structural deficits of the colonial status that increase damages. Danielle Zoe Rivera notes "continuously-poor disaster preplanning and response is a mechanism of deepening coloniality of Being" (2022, 129).

The effects and aftereffects of Maria converge in pragmatic and symbolic interpretations of disaster. Bad Bunny links hurricanes and nationhood in his song "Una velita" tying Maria to climate anxiety and national resilience. He references "Temporal, temporal" and Lola Rodríguez de Tió, with an overlay of climate change.

Witnessing and telling stories of Maria and its aftermath makes clear the greater resilience of communities. First-person accounts of Maria's destruction are affirmative testimonies that resonate with the theme of grassroots resilience, as if people were saying, "We lived through this, we survived." The long Puerto Rican tradition of citizen complaints about governmental inefficiencies now has an even firmer grip on the culture. After Maria protests critiqued notions of "natural" disasters that assign blame to Nature, instead of to the State; for instance, the motto of Colectiva Feminista en Construcción is: "El desastre no es natural" ("disaster is not natural"; Zambrana 2021, 76).

Popularly known as simply Maria, the cyclone made landfall as Category 4, with maximum sustained winds of 155 mph. NOAA's storm webpage reported its many impacts:

> strong winds spread large amounts of debris across the entire area. All full trees were defoliated, and those that were not, were snapped or uprooted by Maria's strong winds and lost medium to large branches . . . Numerous

> locals reported that they felt the ground and their houses shaking, while most were amazed by the force of the unprecedented strong winds that not only transformed their surroundings but also their lives. Although most structures across the island are built using concrete as the main material, countless homes and buildings sustained some type of structural damage. If not blown off, non-concrete roofs suffered some type of damage. Nearly all commercial signs, fences, and canopies were destroyed, including large digital high-definition boards.[2]

These disastrous effects were undoubtedly traumatic for the families that lost a loved one, and those who lacked water and food for their children, or whose house was utterly destroyed. People shared their experiences and offered advice on social media (or as soon as the phone grids operated again), which was how information on government inefficiencies was broadly shared, especially reports of water and supplies outrageously going to waste (Meléndez-Badillo 2024, 222–223).

The archipelago's main infrastructures collapsed: electrical, telecommunications, water systems, and transportation. The failure of these interlinked flow systems proved that they are what Brian Larkin calls "massive infrastructural networks that have come to organize everyday life" (2003, 330). The multiple collapses brought to the surface social poisons such as educational and healthcare deficits, gender violence, an aging population, and depopulation. Speaking of these socio-economic inequalities historian Jorell Meléndez-Badillo says, "hay muchos Puerto Ricos. Y la gente experimentó el huracán María de acuerdo al Puerto Rico en que vivía" ("there are many Puerto Ricos. And people experienced hurricane Maria according to the Puerto Rico they were living in" 2024, 213). Those who lacked adequate help or information felt the trauma of Maria even more, like the deaf community (Marrero Cabán 2022). We have known since the 1990s how people's suffering deepens after disasters; studies show that "about one in four people develop PTSD after a major disaster" (Price Lee 1996, 54). Since trauma is different for everybody, every writer and artist we see in this chapter interprets the experience differently, but using shared themes: grief and mourning, focus on community, and testimonial narratives.

Documented earthquakes have been a factor Puerto Rico since the beginning of the region's written history, happening in: 1615, 1751, 1772, 1780, 1787 (which merited a painting by Campeche), 1851, 1867, and tremors accompanied San Ciriaco in 1899. The earthquakes of 2020 were a reminder that unpredictable geological disasters are part of Puerto Rico's environmental picture, imbricated with long-standing economic and political issues that result in multiple infrastructure deficits. As a result, Antonio Martorell in 2024 shows the map of the island fragmented in the heartbreaking pun "La Isla en cantos" ("The Island in Pieces").[3] Historian Jorell Meléndez-Badillo, who woke up with the first earthquake, documents how unreliable government help had become in 2020 and how communities banded together to help one another (2024, 231–234).[4] Lorna Gisela Jaramillo Nieves memorialized the 1918 earthquake and tsunami in a book. Another, newer Capitalocene condition for disaster is the displacement of Saharan sands which cause droughts (Deopersad and Azevedo). Maria, 2020 earthquakes and the Covid-19 epidemic conform to what Timothy Clark calls "derangements of scale" because of the extreme and extended consequences to the Puerto Rican Nation and environment (2012, 149).

Socio-Political Context

The SCOTUS decision in Puerto Rico v. Sanchez Valle (2016) clarified Puerto Rico's status as a territory, eliminating the political fiction of a self-governing commonwealth. Writing for the majority, Justice Elena Kagan said that "the ultimate source of Puerto Rico's prosecutorial power is the federal government... the Commonwealth and the United States are not separate sovereigns." Trapped in this colonial situation Puerto Rico is at the mercy of federal agencies when disaster strikes. Hurricane Katrina had already demonstrated U.S. federal agencies' disregard for minorities who additionally were, "stigmatized by pejorative media coverage that portrayed them as violent, hapless, or otherwise deviant others—themes that were reprised in Puerto Rico in 2017 during Hurricane Maria's sadly protracted aftermath" (Kierner 2019, 9). Recent disasters remind Puerto Ricans of governmental deficiencies despite it being an American territory. The pervasive feeling of abandonment has led to "communities in crisis [that] benefit more from grassroots relief programs, which can be more effective and less wasteful than their overly bureaucratized official counterparts" (Kierner 2019, 10).

Instances of mismanagement and large-scale corruption in disaster management have been a feature, not a bug, of Puerto Rico's colonial situation well before 2017. Hurricanes and earthquakes make the intermeshed politics of disaster more visible; for example, between 1999 and

2002 five NPP and two PDP mayors were convicted of "extorting from or conspiring with contractors to appropriate almost $22.6 million by billing FEMA for services not rendered" (Duany 1991, 88). The Trump administration used local corruption to excuse withholding many of the billions that Congress assigned for post-Maria recovery.[5] Under the umbrella Act 60 Puerto Rico has offered since 2012 investors tax incentives intended to spur economic activity but which have not brought in the promised trickle-down financial well-being for all Puerto Ricans. A housing crisis, however, has resulted, destabilizing Puerto Rico's economy and morale. These sequential disasters highlight the Unites States' neglectful oversight and the island's home-grown mismanagement. For example, in 2021 earthquakes inspections of damaged schools, some severely affected, had not even reached 69% completion (Caro González 2021).

Patricia Yaeger's term "ecocriticism$" pinpoints how capitalism and colonialism's malignant forces convergence. Yaeger notes that "the ocean as *oikos* or home rolls under, beneath, and inside the edicts of state and free market capitalism. We've left the possibility of wilderness or pastoral for the roller coaster of capital" (2010, 529). The Capitalocene is in full view when hurricanes and earthquakes damage Puerto Rico's infrastructure and environment. Hurricanes also exacerbate the problem of coastal erosion, although in a hopeful event in 2023 the town of Loíza completed a 91-meter breakwater in the Parcelas Suárez area which had shown deteriorating coastal conditions.[6]

Activism and Community Organizing After Maria

In response to these circumstances, Puerto Rico's activists, artivists and climate justice advocates encourage decolonial, community-centered actions for "stripping resilience definitions of their prescriptive, functionalist, and ahistorical presumptions" (Borges-Méndez and Caron 2019, 1,940,001–7). The oppressive situation of vulnerable communities is urgent: "our communities [are] disproportionately impacted by the disappearance of plant and animal life, by hurricanes, fires, floods, soil erosion, heat waves, less water, less access to our traditional foods, and non-GMO seeds" (Lara et al. 2023, 154). Decolonization would involve society-wide, multi-pronged efforts to undo or mitigate ingrained systems of colonial education, development and, for our purposes, disaster aid. Decolonial theory posits that the pervasive violence of colonialism is entrenched in racist, patriarchal, and capitalist extraction systems. Current solutions center on community-led initiatives to wring desired outcomes from the nation's socio-political systems.

Climate justice undergirds the renewed focus on communities. Cultural works in turn emphasize how capitalist systems deepen climate crises and how decision-making and mitigation in environmental matters must involve the most vulnerable communities. This new community sustainability rhetoric is in marked contrast to the anodyne governmental and commercial slogan "Puerto Rico se levanta" ("Puerto Rico Rises").

Puerto Rico already had a history of community organizing after disaster; for instance, after Hurricane Hugo there were calls for, "the implementation of a comprehensive development plan. This included the construction of housing projects where residents whose homes were on unsafe land were relocated, improvements in infrastructure in the sector, as well as a wide range of social, cultural and economic self-management programs" (Naranjo Orovio 2023, 138). Casa Pueblo is among the most lauded of community foundations, a self-managed project for natural, cultural, and human resources. Formed in 1980, it protects acres of land in the municipalities of Adjuntas, Utuado, Lares, and Jayuya. Its director stated on their Facebook page that "5 years after Maria, with limited resources –without a penny of federal funds, without a penny from the [Puerto Rican] government and with an effort from philanthropy– we have been able to transform the energy landscape. The community owns the infrastructure with a microgrid in which we have seen that houses save up to $40 per month."

After Maria there were many DIY community cleanup organizations. The documentary "We Still Here" follows a youth group in the town of Comerío as they aid their neighbors and town. Woman-centered organizations like Fundación Bucarabón, a 501 c(3) non-profit organization founded in 2018, support local development and community organization in Maricao and surrounding areas.

Energy insecurity continues to worry Puerto Ricans. Josh Dzieza explains unequal energy generation: "[m]ore than five years after Maria and despite billions of dollars in allocated federal recovery funding, Puerto Rico's electric system remains in a state of protracted disaster, its 30,000 miles of fragile power lines and antiquated oil-burning power plants plagued by regular outages and at the mercy of surging fuel prices" (2022). Anthropologist Hilda Lloréns adds that "the battle against the impacts of fossil fuel generation in Puerto Rico's southeast is the archipelago's most significant environmental terrain of struggle"

(2021, 96). Thus, community and individual solar energy projects became pressing stories after Maria, in order to counter the U.S. and the local government's trust in upstart, inefficient for-profit energy agencies like LUMA and to move away from petroleum-based energy generation. Many people and communities aspire to have solar microgrids and more resilient solutions to energy precarity. An electric cooperative, Cooperativa Hidroeléctrica de la Montaña, states that: "[t]here are many such initiatives, from community microgrids like Castañer to solar-powered water filtration systems, emergency solar arrays for medical devices, and the 40,000 individuals who have gone out and bought panels and batteries themselves" (Dzieza 2022). The high cost of whole-house batteries, like Tesla's, is one of many stumbling blocks. A community center under the umbrella of organization IDEBAJO (Iniciativa de Ecodesarrollo de Bahia de Jobos) benefitted from funds and grants from private and non-profit organizations for solar arrays and Tesla batteries and now operates completely off the grid (Lloréns 2021, 149).[7]

Culebra and Vieques—this last one famous for its environmental and political protests which led to the closing of a U.S. Navy training base—have become focal points for energy resilience despite their distance from the main island. Vieques's unstable electrical system depends on a submarine cable to transmit power from the main island (also alarmingly, the town lacks a clinic). Much has been written about communities transitioning to solar-powered electrical microgrids. In late 2023 people want a "decentralized grid, one that is more stable and more resilient and powered mostly by the sun" (Bentley 2023). The successes of the Casa Pueblo microgrid in Adjuntas and the Castañer Microgrid in Lares prove sustainable energy is possible. If well implemented, the $1 billion allocated for Puerto Rican energy resilience by President Biden in 2022 could replicate the results of organizations such as these. For example, there is hope that through community-based efforts Culebra could become the first all-solar island in the hemisphere. Energy change in Puerto Rico, however, is ever elusive; Benjamin Torres Gotay tweeted on Aug. 2023 that even though local law mandates 20% of energy be renewable only 4% was. Adam Trexler and Adeline Johns-Putra state that solutions to climate change "require network and negotiation, not magic bullets nor heroes" (2011, 2). Additionally, as Cynthia A. Kierner explains, it remains true that state disaster benevolence is like the Anglo-American world of the late eighteenth century, "the performance of benevolence was also a tool of statecraft, a political tactic deployed to mitigate colonial discontent, strengthen the imperial bond, and solidify a shared sense of British national identity" (2015, 100). Plus ça change.

Local governance often prioritizes capital development over community well-being. It says something about national priorities that a new shopping center opened in January, 2021, even before the availability of the Covid-19 vaccine. In the summer of 2022, five years out from the hurricane Maria grid debacle, Governor Pierluisi announced that it would take eight years for the electrical grid to be fixed. Local incompetence was evident when he announced that FEMA had approved only 15 of 219 contracts, representing only 1% of the total congressionally-approved funds for new electrical grids.

Given these many deficits and hopes, Maria became a cultural milestone as stories, images, and symbols processed the extended deprivations of daily life and failed infrastructures that caused suffering and increased migration.

Like a cultural tempest, a supercharged Puerto Rican culture has continued interpreting disaster as a metanarrative. Literature saw a quick proliferation of testimonial narratives. Art inspiration came from scientific data and a landscape of debris. Not since San Felipe II in 1928 has the culture focused so much on disaster as an agent in the national story. For example, current historical writing deals with the climate more than ever. Luis Caldera Ortíz's 2017 *La historia de los ciclones y huracanes tropicales en Puerto Rico* ("The History of Tropical Cyclones and Hurricanes in Puerto Rico") offers detailed hurricane data from the sixteenth century onward and also the local effects of global climate change.

Songs, cartoons, novels, plays, films, documentaries, monuments, museum exhibits, and more have memorialized and documented how Maria changed Puerto Rican life. For instance, founded after Maria, the journal *Categoría 5* renews calls for political independence as the solution to the ills that the storm exposed: the harm of archaic shipping laws, the indifference and incompetence of local and federal agencies, a deteriorated infrastructure. In addition, Puerto Rico's grinding national debt "actualizes, updates, reinstalls the colonial condition" (Zambrana 2021, 29). Culture works to process, memorialize, and record people's shock, despair and grief, as well as the collective support of communities.

Ana María Fuster Lavín's micro-story "Arrojados al silencio" ("Flung into Silence"), from Nov. 2, 2017, is a brilliant example of material feminism with its representation of death, silence, and darkness after Maria for both humans and non-humans. The narrator regards three baby bats flung to their death by the storm as "niños-murciélagos" (chidren-bats"), stressing how all living beings were fragile during the hurricane.

Two of Mayra Santos-Febres's books bear witness to storm life: a poetry volume aptly titled *Huracanada* ("Hurricaned," 2018) and a memoir, *Antes que llegue la luz* ("Before the Light Returns," 2021). She had already used a storm strategically in her novel *Nuestra Señora de la Noche* ("Our Lady of the Night," 2006); Charlotte Rogers stresses that Santos-Febres fixes Isabel Luberza's birthdate on the date of San Ciriaco (2020, 185). Cultural legacies about hurricanes cast long shadows; one of Santos-Febres's poems recovers and transforms José de Diego's image of the storm as an enemy to be fought in his popular poem, "En la brecha" ("Unto the Breach") from 1904, five years after the monstrous San Ciriaco. Santos-Febres's poem takes the Androcene trope of Man overcoming storm and flips it to a community of women who, machetes in hand, fight Maria's multiple crises and give symbolic birth to a new nation. Santos-Febres seeks to "better understand the emotionality of the resources that exist in everyday struggles" (Sultana 2011, 163) and indeed emotions are at the heart of the poem, which also references Lola Rodríguez de Tió's poem "La borinqueña" in which machetes are the weapon of choice. Hilda Lloréns says the hurricane's aftermath imposed an "exponential overburden on women" (2021, 126) and Santos-Febres's women take on that burden when they clear out debris to reconstruct a ruined landscape:

> Tenemos las manos sucias, / el pelo lleno de arena, / las espaldas truncas, el pecho en llanto, / pero hemos sobrevivido a la furia del viento. / Salimos guerreras, / machete en mano, hijos a cuestas, / a buscar alimentos, agua, a desclavar tormenteras, / a destapar alcantarillas. Junto a otras, laboramos. Quiero que se sepa, / tomen nota. / Podemos, entre todas / levantar lo caído, parirnos una isla nueva. ("Our hands are dirty, / our hair full of sand, / our backs truncated, our breath in tears, / but we have survived the fury of the wind. / We come out women warriors, / machete in hand, children in tow, / to look for food, water, to unnail shutters, / to unclog sewers. / Together with others, we work. / I want it to be known, / take note. / We can, among all of us / raise what has fallen, / give birth to a new island"; 2018, 61–62)

The poem offers post-Maria themes: testimony, community action, focus on preparedness, the ways of resilience. Víctor Figueroa notes Santos-Febres play on "dirty," distinguishing between the dirty power games of the powerful and the "dirty hands" of workers who aid recovery (2018, 13). Its communal heroes supplant an unreliable government in order to overcome disaster: "Todos salimos a la calle / armados tan solo con nuestras manos / a arañar la destrucción" ("We went out into the street / armed only with our hands / scratching at destruction"; 2018, 52). Ecocritic Adeline Johns-Putra stresses the importance of placing a text "in its collective cultural, ethical, emotional moment" (Yuan 2022, 8); for Santos-Febres the moment of collective unity came immediately after Maria because, as she says, "[s]e cayeron todos los gobiernos locales, pero la gente se puso de pie" ("all local governments fell, but the people stood up"; Laborde 2018). The image of impromptu brigades clearing environmental devastation stresses that community action sprouted everywhere.

Huracanada reaffirms the hopeful belief in natural succession, that the non-human world renews: "Ya regresan los colibríes a libar de las flores / del roble que quedaron abiertas / después de la tormenta" ("The hummingbirds are back to sip from the flowers / that were left open in the oak / after the storm"; 2018, 73). Accordingly, her trust in natural succession is at it was in Abbad's time and indeed, Jeannine Murray-Román shows how literary hurricane accounts compare to early modern *crónicas* (one is Audre Lorde's another Sofía Gallisá Muriente's poems) (2018, 20).

Santos-Febres casts the hurricane as storyteller: "[e]s el ojo del huracán el que cuenta todo esto... yo tan solo tomo nota" ("[i]t is the eye of the hurricane that tells all this... I just make note"; 2018, 69). Remembrance of the dead, an important post-Maria theme: "Recordemos sus nombres, sus caras" ("Let us remember their names, their faces") and continues in a triumphal nationalist tone, "[n]uestro pueblo se enfrenta a la muerte / y sale airoso" ("our people face death / and prevail" 2018, 64).

The popularity of testimonial accounts and memoirs after Maria rewrites the old kinds of Androcene heroic narratives we saw in earlier centuries, in order to emphasize community-based interventions, bypassing an inept and corrupt colonial system. *Antes que llegue la luz* blends fiction and recollections centered on community collaborations. For instance, she goes to the Colaboratorio (a community incubator of NGOs) to access the internet.

Post-Maria portrayals of the environment represent many such shifts and changes. Bad Bunny's *El apagón* ("The Blackout" from 2022) declares disaster is a way of life. Its accompanying documentary, Bianca Graulau's *Aquí vive gente* ("People Live Here"), traces displacement of property ownership in a system that encouraged profiteering and land grabs after Maria.

There is also a significant change in the color palate of the landscape. Traditional views of the Puerto Rican landscape featured what Margot Arce de Vázquez called (in 1939), "la calidad fosforescente de los verdes vegetales" ("the phosphorescent quality of vegetable greens"; Quintana 1996, 293). The color green, as we saw in Chapter 4, is also associated with nationalism, like with the Partido Independentista Puertorriqueño, the pro-independence political party. In art after Maria blue replaces green, specifically the blue *toldos* dotting the landscape for years afterward, colorful symbols of a stifled recovery. Another environmental motif was gas-powered generators drowning out the song of the traditional *coquí*. Paper towels, infamously thrown by Donald Trump, would also become symbolic of colonial disregard for Puerto Rico.[8]

Anthologies of testimonies and stories about the hurricane as a social, political, and cultural phenomenon appeared as early as 2018. *Bitácora de una transmisión radial* ("Binnacle of a Radio Transmission") constitutes a ship's log of the nation as it battled the storm. The anthology selects transcripts of raw radio transmissions from Cadena Wapa, the only radio chain transmitting continuously during the storm. It compiles interviews with authorities and anguished calls for rescue. The author, Sandra Rodríguez Cotto, highlights a text and photo diary she kept from Sept. 18, 2017, to Dec. 31, 2017, which blends photographic and eyewitness accounts tohighlight experiences during the storm.

Another anthology, *Pa la posteridá. Antología sobre el paso del huracán María por Puerto Rico* ("For Posterity. Anthology on the Passage of Hurricane Maria Through Puerto Rico"; Orsanic 2018) includes archipelagic and diasporic Puerto Rican prose and poems running the gamut of emotions. The environment is described as being radically transformed by, as Eric Rivera Martínez says, "los arpeggios huracanados que desdibujaron el paisaje" ("the hurricane arpeggios that splotched the landscape"; 2018, 101). Most of the pieces focus on the challenging lack of electricity and internet. Writers also focus on many community solutions.

Iris Morales's *Voices from Puerto Rico. Post-Hurricane Maria* centers of diasporic and archipelagic voices united in solidarity as the volume features testimonies of the outpouring of aid, fund-raising and comfort from those two communities united. Morales (who was on the island by October 4th) praises citizen volunteers who helped when government agencies failed. The book is devoted to accounts of life during and after the disaster with snippets of testimony, poetry, stories, mostly from average citizens and members of community organizations. *Voices* thoroughly denounces widespread causes of disaster besides Maria: "massive poverty, lack of infrastructure, and environmental damage resulting from more than 120 years of exploitation and extraction" (2019, 3). The latter part of the book is devoted to solutions to pressing issues such as food insecurity.[9]

Electricity became more central to depictions of the landscape after the grid's failure. In a striking image, Luis Rafael Sánchez presents electricity as the defining feature of the landscape. No slouch when it comes to a point of view inside an airplane in "La guagua aérea" ("The Air Bus"), Sánchez begins *El corazón frente al mar* ("The Heart Before the Sea"; 2021) with an airplane view descending to San Juan at nighttime. He destabilizes the traditional garden island image with: "un fantástico jardín nocturno. El regalo primero que le aguarda a quien llega durante la noche a la capital puertorriqueña estriba en un ramo cundido de flores eléctricas" ("a fantastical night garden. The first gift that awaits those who arrive overnight in the Puerto Rican capital lies in a bouquet full of electric flowers"; 19). Sánchez then reverses the image as the traveler arrives during daytime to see, "un fantástico jardín de piedra y zinc" ("a fantastical garden of stone and zinc"; 2021, 23), again upending the traditional garden allegory. For Sánchez the island's oceanic setting, its address if you will, is in both a lovely and traitorous geographical and historical situation because the seas bring disaster: "[s]obre la piel voluble de esos mares, igualmente cariciosos y traidores, navega buena parte de la historia puertorriqueña... porque la menor de las Antillas Mayores ubica en el paraje donde convergen la avenida Atlántico y la esquina Caribe" ("on the fickle skin of those seas, equally caressing and traitorous, sails much of Puerto Rican history... because the smallest of the Greater Antilles is located in the place where Atlantic Avenue and Caribbean Corner converge"; 2021, 21). Sánchez responds to the same impulse as Caribbean creators who "have long theorized the ocean in terms of the violent convergence of environment and history" (Deloughrey 2017, 33).

Logically after 2017 there would be post-apocalyptic stories; such is Xavier Navarro Aquino's novel *Velorio* ("Wake," 2022). Themes of memory and grief already appear in the title, a reference to Francisco Oller's masterwork *El velorio*, which is set during a funerary rite, a *baquiné* for a dead child. The novel takes place after the passing of "ella" ("she") or "la monstrua" ("the [female] monster"), since Maria needs no naming. In this post-apocalyptic land, after governments have failed, a homophobic and ageist cult leader called Urayoán rises and violently establishes a new dystopian town ironically called Memoria. The most grim storyline is Camila carrying the rotting corpse of her sister, who died in a mudslide; a symbol of the tenacity of grief and necessity of remembrance. The dead sister in this macabre novelistic *baquiné* narrates the novel's last section.

Velorio indicts the inefficiency of local and federal governments, as is clear when Urayoán throws paper towels at his followers with Trumpian disregard. Even so, the belief in natural succession persists even in this novel: "[l]a naturaleza se adapta sin miramientos. Lo digo porque no hay nada como nuestros bosques tropicales, que se reconstruyen por sí solos.... El modo en que un flamboyán, aún caído, puede adaptarse echando retoños erráticos siempre y cuando algunas raíces permanezcan en la tierra" ("nature adapts unceremoniously. I say this because there is nothing like our tropical forests, which rebuild on their own.... The way in which a flamboyán, still fallen, can adapt by casting erratic saplings as long as some roots remain in the ground"; 2022, 89). Resilience indeed.

Isabel Ramos and Kisha Tikina Burgos's 2023 play *Antígona frente al mar* ("Antigone Before the Sea"), also features the theme of mourning. This tragedy honors Maria's dead ritualistically through Memorias, the chorus, giving voice to national remembrance and grief. Like a bell tolling, the number of the dead is repeated throughout the play, "Trescientos, cuatro mil, ochocientos catorce" ("Three hundred, four thousand, eight hundred and fourteen"). The chorus also stresses how precarious colonial conditions cause death, especially life in the dark: "También morimos en vida... / Con la ausencia, la penumbra, la precariedad" ("We also die in life... / With absence, gloom, precariousness," 2023). The leader, Creon, has condemned the nation to the darkness of misremembering, so the nameless Madre in search of her dead son says to the tyrant, "en esta oscuridad a la que tú nos has condenado, no hemos podido encontrar a nuestros hijos" ("in this darkness to which you have condemned us, we have been unable to find our children," 2023). Darkness is forgetting.

Like Luis Rafael Sánchez in *La pasión según Antigona Pérez* ("The Passion According to Antigone Pérez," 1968), Ramos and Burgos update the myth of Antigone to honor the dead from the 2017 storm but also to remember past injustices, like the 1978 Cerro Maravilla killings; here Antigone's dead brothers are killed in a state ambush as they redistribute hurricane aid. Even though both brothers were equally rebellious, the tyrant arbitrarily chooses their symbolic value, one as hero and the other as villain.

Elidio La Torre Lagares processes "wounded landscapes" (2019, 18) with mournful and angry post-Maria poetry that concentrates on the assault on the senses of the passing hurricane. A poem called "natural disaster number #1: there will be mud," features mudslides that bury people. The sense of dread is palpable: "the air sours with putridness / amid the tailing whips of the hurricane / my heart pounds against the dark sky / as a wave of dead birds rakes the pavement / my breath turns to ashes / in the soliloquy of the wind" (2019, 3). The remembrance of loss is pervasive, and the grimmest poem is titled "baquiné for Puerto Rico" connecting again to Oller and themes of mourning, like Navarro Aquino, and Ramos and Burgos. La Torre Lagares projects his ire beyond the local and federal governments, singling out Elon Musk—the poster boy of opportunistic late capitalism—for his empty boast that he could power up Puerto Rico which, needless to say, he never did.

Post-Maria's seismic migration to the U.S. impacted daily life as well as culture.

This migration caused a dramatic population drop: "more than 135,000 Puerto Ricans relocated to the United States six months after Hurricane Maria" (Meléndez 2018). The depopulation process had started with 2006's financial crisis but migration skyrocketed after the hurricane, heightening existing social crises and severely straining resources. In 2021 *NotiUno* reported the many detrimental ways that the population decrease has affected inland towns,[10] with a flood of young workers leaving, displacing key health care workers and leaving an aging population with diminishing resources.[11]

Tere Dávila features migration in her short story "Génesis" (2021) which uses apocalyptic references and a Biblical style. The storm lasts seven days, and the blackout lasts forty days and forty nights:

> Y sucedió que al próximo día los vientos vinieron sobre la tierra, se abrieron las fuentes del abismo y las cataratas de los cielos.... Entonces llovió siete

> días corridos y las gentes se lamentaban porque una gran oscuridad cayó sobre la tierra. El apagón ya llevaba cuarenta días y cuarenta noches.... Para colmo, Cholo estaba sin incinerar porque la funeraria no tenía generador. Irkamary...escapó del barrio con una de las doñas de la Atalaya y nunca miró tras de sí, ni siquiera por la ventana del avión que iba rumbo a Florida. ("And it came to pass that the next day winds came upon the earth, the fountains of the abyss and the cataracts of heaven opened.... Then it rained seven days in a row and the people lamented because a great darkness fell on the earth. The blackout had already been going on for forty days and forty nights.... To top it off, Cholo was not cremated because the funeral home had no generator. Irkamary...escaped the neighborhood with one of the Watchtower ladies and never looked back, not even through the window of the plane bound for Florida")

Dávila's apocalyptic tone reflects the real-life feeling of end times of many Puerto Ricans after Maria. The protagonist migrates to Florida, which has replaced earlier migration foci like New York and New Jersey (2018).

The bonds of island-bound and diasporic communities run deep. It is beyond the scope of *Disaster Nation* to look at all diasporic cultural products that refer to Maria; we sample some that highlight our disaster themes. Dorsía Smith Silva's *Inheritance of Drowning* (2024) has drowning as central allegory not only in her vivid poems about the storm but also referring to the many agonizing drownings of life in a colony during the Capitalocene: "we can't escape the drownings / determined to handcuff / us below sea level" (33). The diasporic poet shapes her perceptions of the before, during, and after of hurricane Maria building connections to her personal and national identities using debris as the symbol of multiple conditions of disaster.

Esmeralda Santiago joined the cultural explosion of post-Maria narratives with *Las madres* ("The Mothers"), a woman-centered melodrama of five women whose stories switch between the 1970s and 2017. The women's impending trip to Puerto Rico is canceled when Hurricane Irma approaches Puerto Rico but, fatefully, they reschedule their trip and become trapped in Maria. *Las madres* shows blow-by-blow the anxiety and suffering of the storm's passing as well as the scarcity of resources afterward. Harrowingly, one of the women, a nurse, cares for people who have no other recourse for medical treatment. *Las madres* has an extended section of minute details of storm experiences, listing deaths, and the devastation in many municipalities (2023, 291–297).

Another diasporic writer, E.G. Condé, opens their speculative fiction novel *Sordidez* ("Squalor") with a hurricane. In a dystopian future where the U.S. has ceded control of Puerto Rico to China, one of the lead characters describes a storm named Teddy (in honor of Teddy Roosevelt) personifying the cyclone as a monster, as we have seen since Rafael Tufiño's *La Plena*: "I was huddled with my family when his tongues slithered through cracks in our zinc roof, bored holes in the tarp that we hoped would shield us." The storm then enters with "jaws bared, his saliva oozing down on us in wicked rivulets" (2023). The novel relies heavily on Taino mythology, for example, a character turns into Jurakán, and Guabancex is a ferocious, vengeful character. The author's didactic postscript is a call for the island's independence and a warning against climate change.

Migration is equally stressed as a theme post-Maria. Nadia Hallgren's documentary *After Maria* (2019) depicts the shared misery of Puerto Rican climate refugees, three women and their families displaced to a FEMA motel in the Bronx. One of the women characterizes the storm in ways seen before: she feminizes the monster storm, "Ella llegó como un monstruo barriendo completamente a Puerto Rico" ("She came like a monster completely sweeping over Puerto Rico"; 2023). Hilda Lloréns says it "exposes the hardships they face, but also the caring and sustaining ways in which disadvantaged women forced to live on the capitalist margins try to provide dignified lives for their families" (2023, 136). *After Maria* was controversial on social media with bitter arguments between those who stayed vs. those who left. Many people in Puerto Rico saw the former as hard-working, patriotic people vs. the latter as lazy, unpatriotic women who walked away from their homeland.

Xavier Valcárcel's chronicle and testimony *Aterrizar no es regreso* ("Landing Is Not Returning") recounts his experiences during the hurricane, his feelings as a climate refugee in New York, and his emotional return to the island. The hurricane represents an awakening: "el huracán sacó a todos de su comodidad. Arrancó las máscaras al país, voló las gríngolas, y nos vimos como siempre fuimos, aislados, a la merced de lo que trajera el mar o el aire" ("the hurricane took everyone out of their comfort zone. It tore off the masks of the country, blew the fetters, and we saw ourselves as we always were, isolated, at the mercy of what sea or air would bring"; 2019, 58).

Edgardo Miranda-Rodríguez's graphic tales pre- and post-Maria, *La Borinqueña* (2016) and its companion piece *Ricanstruction* (which he produced, 2018), view Puerto Rico and the diaspora through the lens of disaster. As Jossiana Arroyo says, *La Borinqueña* "raised the Puerto Rican archipelago and its natural and man-made ecological disasters as key concerns for cartoonists and graphic artists" (2023, 139). *Ricanstruction* presents an ecocritical history of Puerto Rican disaster, depicting the metanarrative of disasters. The first section is set during 1928's San Felipe as people survive out in the *campo*, later a section on hurricane San Ciprián brings disaster continuity. The narratives defend reconstruction through community work.

Both of these graphic anthologies feature an Afro-Puerto Rican diasporic superhero and explore "a much longer narrative arc of anticolonial resistance" (Humphrey 2021, 67). The root of superhero Marisol's (La Borinqueña) power is her heritage as she gains: "the powers to control the wind, water, and land from the close relationship she develops with the Indigenous Taino spirits and people of Borikén" (Humphrey 2021, 67). Most of the segments connect to traditional anti-colonial pro-independence icons: Lares, Pedro Albizu Campos and Taino cultural legacies, such as Jurakán. Direct references to Puerto Rican culture abound; some segments enact what Frederick Luis Aldama says Latinx superheroes always do: challenge racial and other stereotypes (2017, 174). The comic from 2016 is also prescient about the island's failing electrical infrastructure post-disaster.

The follow-up anthology, *Ricanstruction. Reminiscing & Rebuilding Puerto Rico*, is specifically about hurricane Maria and its aftermath. La Borinqueña delivers the earnest slogan, "el pueblo unido jamás será vencido" ("a united people will never be defeated"; 2018, 38). The diaspora community indeed rallied to help their brethren, side-stepping the government; Lin-Manuel Miranda (to whom a section of *Ricanstruction* is dedicated) composed a fund-raising song, along the traditional lines of "We Are the World"—called "Isla Bendita-Unidos Por Puerto Rico" ("Blessed Island—United for Puerto Rico") including achipelagic and diasporican artists. *Ricanstruction* also has many narratives of disaster life; for instance, the beginning section, "Living in the Dark for 144 days," has no superheroes, instead a heroic family forages for basic necessities during the extended blackout. The mother recounts the many challenges with no electricity, including mosquitoes, always a dreaded tropical health hazard (spreading dengue, chikungunya, Zika), which became a fierce enemy during the blackout. Traditional narratives of the (home)Land did not feature mosquitoes, but after Maria stories and art often feature them as a constant presence in the landscape.

A prominent theme in the anthology is how not superheroes but people saved people in community solidarity: "María took a lot away from us, but it also revealed the spirit of solidarity in Puerto Ricans" (2018, 7). *Ricanstruction* highlights the community resilience of regular folks shouldering the load of rescue and reconstruction. For example, a segment about a little girl who sells Puerto Rican delicacies to raise money for reconstruction. In order to help Puerto Rico the anthology also included a URL for donations.

Ricanstruction also highlighted the mass protests which became part of the Puerto Rican political scene in 2019 following RickyLeaks a political scandal involving then-governor Ricardo Rosselló, which led to his resignation. We see a Puerto Rican timeline of colonial protest against government from the anonymous letter of 1679 (see Chapter 2) to the intense *perreo* protests in front of the San Juan cathedral.[12]

Non-humans play an important part with the focus on the devastation of iconic natural environments on the island, like *El Yunque*. Traditional fauna are leading characters in "Song of Coquí" where a local parrot, a traditional symbol of an authentic environment, saves the tiny melodic frog, another symbol of the Nation. After decades of recovery efforts Maria, sadly, decimated the imperiled species of parrots. Because of the anthology's optimistic tone, the *coquíes* also produce medicine. The anthology uses iconic national non-humans: the *coquí*, the *cotorra* and the *pitirre*, which connects back to José de Diego' inspirational poem about a resilient and combative *pitirre*, like Puerto Rico: "that tenacious bird is who we are inside... It's who we strive to be against all odds" (2018, 91).

Multiple references to Puerto Rican culture also ground the anthology in the *patria*. Luis Palés Matos stars in a time travel section. These graphic stories of interwoven Puerto Rican elements brightened the collection's sunny, positive outlook, even when noting the Jones Act damages.

As mentioned before, blue *toldos* emerged as powerful cultural symbols post-2017 as artistic uses of toldos were creative indictments of the Trump administration's lack of efficient dispersal of the funds appropriated by Congress for Puerto Rican aid and recovery (Fig. 5.2).

Jotham Malavé's oil painting *El bautismo colectivo* ("Collective baptism," from 2020), shows a blue *toldo* covering naked people; the *toldo* a blinds them symbolizing disaster dependency and neglect. In 2021 the

Fig. 5.2 Jotham Malavé. *El Bautismo Colectivo*. Oil on canvas, 2020. Instituto de Cultura Puertorriqueña Collection

toldo was also the inspired choice of a Yara Sofía, a Puerto Rican contestant in *RuPaul's Drag Race All Stars* who se gown was made of blue *toldos*.

Storm debris became powerfully significant, sometimes literally as found objects and always symbolizing a detrital colonial situation. MacArthur Fellow Daniel Lind-Ramos's sculptures *Tres Marías* and *El Viejo Griot* use disaster wreckage. Rubén Colón Morales, in a column titled "Un país de escrombros" ("A Country of Debris") calls on Puerto Ricans to choose between loss or self-recovery. The theme of self-reliance also appeared in an anonymous viral image of children clearing debris from a road around the time Donald Trump tweeted that Puerto Rican people "want everything done for them." The children's impromptu cleanup brigade reaffirmed grassroots strength, community, and collective action, showing that even the young and innocent can clear the way to the future.

Arts Pre- and Post-Maria

Perhaps the most striking development regarding the topic of this book, how culture interprets the environment and the metanarrative of disasters, is the post-Maria sea change in the plastic arts from traditionally serene landscapes to hurricane-centered images and symbols. It did not happen overnight. The first part of the twentieth century depicts the landscape as seen in the nineteenth century. The Museo de Arte de Puerto Rico's online exhibit, *The Memory of the Landscape*, proves Puerto Rican art has had a tradition of landscapes that ecoerase storms. The exhibit has examples from late nineteenth century to the first four decades of the twentieth. María Cadilla's *Paisaje río Bayamón* ("Bayamon River Landscape" ca. 1940) is a flat horizontal view presenting the many colors of a river bank, with iconic *palma real* trees and *flamboyán* trees central to the composition. These two iconic plants are motifs throughout the exhibit. Most of the artworks, for instance Gonzalo Gil de León's *Paisaje campestre* ("Country Landscape" 1923), put houses in the middle ground of the composition, as the human element takes precedence over the non-human. Rafael Tufiño's *Sin título* ("Untitled") centers on a people sitting outdoors. Social commentary is never far from Puerto Rican art, Margot Ferra's *Fanguito* shows poverty in the landscape. Yet no artwork in the exhibit shows storms (or even bad weather) as the tropes of eternal spring and garden island remained strong into the 1940s.

Twentieth-century Puerto Rican art moved toward realistic representations of disaster, and grew past traditional ecoerasing landscapes, until it finally opened up in rapprochement to imagery of hurricanes. Chapter 4 explored artworks featuring hurricanes starting in the 1950s Carlos Raquel Rivera's print *Huracán del norte*, José M. Figueroa's poster *Huracán*, and Eduardo Vera Cortés poster *Huracán*, as well as Tufiño's monumental *La plena*. These works elevated the hurricane to a main character presiding over the nation. It is true that significant works of art still failed to engage the environment (which we also saw with Lorenzo Homar) but ecoerasure became increasingly untenable when the twentieth century ended.

Representations of disaster increased significantly in the twenty-first century. In 2012, Frances Gallardo began her ecopoetic *Hurricane* series highlighting the delicate balance in local ecosystems. Using cut-out paper and cross stitch embroidery, her swirly imagery replicates the counter-clockwise rotation of hurricanes. Her delicate choice of media contrasts with the indestructible power of cyclones. Paul Delano's ongoing conceptual installation *The Museum of the Old Colony* (started in 2016) features images, text, and objects going deep into colonialism; includes two disaster photos from 1899, a flooded street scene in Arecibo and a U.S. War Department photo of Yabucoa. These photographic testimonies of San Ciriaco's destruction show people in ravaged landscapes; the latter with a horizon of destroyed houses. Delano traces the disaster metanarrative speaking of the Trump administration's lack of responsiveness: "If you read some of those original captions, they are identical to the tone and content of what Trump was saying" (Gonzalez 2017). Another diasporic tribute to struggles and resilience is Antonio Martorell's *Mi Grito Hacia el Mundo* ("My Cry unto the World"), a sculpture in Battery Park, New York, whose glass panels depict an "ascending spiral [which] is topped off by an upward rotating star representing Puerto Rico supported by five glass walls printed with descending colored streams."[13] Using the colors of the Puerto Rican flag, Martorell remakes Puerto Rican identity, integrating hurricane imagery and patriotism.

In 2017, the Museo de Arte Contemporáneo de Puerto Rico's exhibit, *Trabajo de campo* ("Fieldwork") featured artist Rosaura Rodríguez, who uses elements of the landscape as material for artforms. Made from local botanical and mineral materials, her artworks explore the botanical changes in the green landscape. Paul Humphrey explains how: "Rodríguez produces pigments from natural materials collected on Tabonuco farm, located in Jayuya" (2021, 66). These are also further links to an "authentic" landscape.

Post-Maria the hurricane becomes a character of consequence in Puerto Rican art as newer images show the impacts of the bioregion's violence. As part of Enoc Pérez's 2023 ironically-titled exhibit *Las Islas del Encanto* ("The Isles of Enchantment") at the Museo de Arte de Puerto Rico, the palm tree, one of the most traditional symbols of the island, appears uprooted. We have seen this island icon standing proud and tall since at least Tapia and it certainly features prominently in touristic images. In five separate paintings, however, Pérez knocks the palm tree on its side to show torn-up roots, like life post-Maria.

There is also interest in the rights of the non-human, which in its simplest formulation centers on who and what gets to be a character or protagonist in the stories of the Nation. Does a tree have the same rights as a person? The controversial rights of nature legal issue has not, and may never be settled but it is important for some artists explored in this chapter. Legal personhood for non-human entities has been fought for and sometimes won: New Zealand's Whanganui River, India's Yamuna and Ganges rivers, and the Colombian Amazon. Ecuador topped them all in 2008 when its constitution famously enshrined the rights of "Pachamama" ("Mother Nature"). On December 2018 Toledo, Ohio residents made a case for an emergency "bill of rights" for Lake Erie proposing the ecosystem around the Lake be granted legal personhood, with all the rights of law. Alison Byerly says so-called deep ecologists center the issues of the legality of the environment around concepts of moral standing and rights. She adds that the most radical proponents, the members of the Earth First! movements see, "'[t]he Defense of Mother Nature' as a task that merits the same human risk and civil disobedience as the antislavery and civil rights movements" (1997, 63). The matter of responsibility has complicated legal ramifications; could you sue a river if it floods your property?

The theme of "the rights of nature," centering on the non-human, is central to the Museo de Arte Contemporáneo de Puerto Rico's online exhibit, *El Momento del Yagrumo* ("The Moment of the *Yagrumo*"). It also includes independence-heavy themes of "demands of sovereignty and resistance." The *yagrumo* tree symbolizes rebirth and recovery after a hurricane, since it survives storm winds when stronger trees fall and germinates in unexpected hurricane clearings. The theme of the natural succession has changed because we cannot have blind trust that the environment will self-renew, as this exhibit clearly states, nature's rebirth is momentary until the next storm comes, highlighting "the need for deeper changes in our relationship with nature." The exhibit materials also link storms to collapsing global economies. Among the artworks is Guillermo E. Rodríguez's *Doppler Landscapes* (2019), a "meteorological twist" silkscreen series that artistically interprets scientific forecasting technology since each piece uses Doppler radar images, melding climatological tech, environment and art.

No Existe

A consequential post-Maria art milestone, the Whitney Museum's 2022–23 exhibit *no existe un mundo poshuracán: Puerto Rican Art in the Wake of Hurricane Maria* explores themes of ecology and landscape, infrastructure, grief, resistance, and protest and touristification. This unique moment allowed Puerto Rican and Diasporican artists to engage with the world because for the first time ever the museum featured only Puerto Rican-identifying artists in a wide-ranging sampling of works, most of which had the environment and hurricane Maria as their subject matter. We will sample works from this exhibit which fully engaged with disaster.

Although the exhibit dealt with the metanarrative of disaster, despite its title, other topics not directly related to the hurricane or the environment laid bare the background for Maria, i.e., the socio-political pre-conditions for such a disaster. There were some representations of nationalist tropes, as well as denunciations of colonial neglect and disaster capitalism. The larger story of the disaster nation appears through traditional pro-independence symbols, for instance, Danielle De Jesus's *Google the Ponce Massacre* (2021) (Guerrero 2022, 78–79), a piece about the 1937 massacre tying U.S. colonialism to disaster. Armig Santos' *Procesión en Vieques III* ("Procession in Vieques III") (2022) (Guerrero 2022, 102–103) is about the death of the U.S. Navy base guard David Sanes in 1999 which triggered protests, drew international attention, and culminated in the Navy leaving Vieques. Yiyo Tirado Rivera's *Castillos de arena* ("Sand Castles") (2020) (Guerrero 2022, 110–111), indicts the house of cards that is the tourism industry, as a replica of La Concha hotel made from beach sand symbolizes the deep capitalist hold on the nation. Hilda Lloréns says that hotels specifically contribute to environmental degradation: "In the Caribbean, hotels are major contributors to greenhouse gas emissions" (2023, 142). Tirado Rivera symbolically places sea sand as a victim of Capitalocene depredations, which Elizabeth DeLoughrey and Tatiana Flores affirm are: "a Caribbean experience of the sea arising from traumatic histories of crossing, state violence against migrants at sea, and the tourism-driven alienation of local people from their coasts" (2020, 165). Patricia Yaeger says tourism degrades the environment by relying on "the recreational sea; the industry cleans up some tourist hot spots and degrades others" (2010, 539). Sofía Gallisá Muriente's video *collage B-Roll* denounces land exploitation through its satirical use of promotional footage taken from Puerto Rico Tourism Company videos made to attract foreign investment. For Gallisá Muriente the tourist industry is directly connected to capitalist extraction and how the real estate speculations that tax haven Act 60 precipitated through speculative investments (many from the U.S.) increased housing insecurity in Puerto Rico.

Gabriela Salazar's *Reclamation* (2022) (Guerrero 2022, 100) could be in remembrance of hurricanes San Roque II in 1893, San Ciriaco in 1899 and most definitely 1928's San Felipe II, which decimated the island's coffee industry on the island. Salazar addresses the coffee industry in relation to disaster and features a blue covering reminiscent of the *toldos*, thus directing the viewers' attention back to Maria. Raquel Salas Rivera's book, *while they sleep[1] [1] under the bed is another country*, gave the exhibit its title It juxtapos post-Maria trauma in poems and footnotes, intensifying notions of colonial "trauma as in centuries" (2019, n.p.). The poem "that time we existed because we were dying" and its footnote, "abrió Plaza Las Américas" ("Plaza Las Américas opened"; 2019, n.p.), notes the persistence of consumerism in the face of disaster. Other after Maria life conditions in the exhibit are: Trump, blackouts, noisy generators, relief loans, lack of water, food scarcity, and airplane tickets. One simply says "help" with a footnote of "filas" ("lines"; 2019, n.p.). Salas Rivera juxtaposes the myth of the endless sea, "a whole ocean," with "la playa está destruida" ("the beach is destroyed"; 2019, n.p.).

Projected on the largest surface at the entrance to the exhibit, Sofía Córdova's video (105 min.) *dawn_chorus ii: el niágara en bicicleta* ("dawn_chorus, ii: Niagara on a Bicycle," (2018) (Guerrero 2022, 74–77), consists of many humorous testimonies of Maria's tragedies tinged with colonial farce: "There's a hospital ship waiting for sick Puerto Ricans–but no one knows how to get on it."[14] Córdova's humor is explicit; a woman called Veronica says people prayed as they left Walmart before the storm, hoping their prayers would help—a religious belief seen since the seventeenth century. She talks amusingly about the shortages after the storm: "No había ni Pepsi ni Coca-Cola. Imagínate" ("There wasn't even Pepsi or Coca-Cola. Can you imagine").

The video's testimonies of disaster also contrast the storm and its debris with the garden myth directly. A direct reference to it shows footage of horses a in lush land while an old song plays: "Puerto Rico es la tierra bendita, la isla del mar y el cielo... bendecida por mi Dios" ("Puerto Rico is the blessed land, the island of sea and sky... blessed by my God"). In another segment a woman walks in a lush, green forest path and ends up by her debris-filled house. Her litany of woes about daily life after

Maria is compelling: the heat, the mosquitoes, the anguish, the fear of the dark night, the extensive lines at supermarkets and gas stations. Córdova's interviewees stress daily anguish as everyday-life turned nightmarish when chores like getting mail or washing clothes seemed impossible. Many, many shots in the dark submerge the viewer as in the months-long blackout: a tattered national flag in the dark; flooded streets; petroglyphs in the dark. Córdova's video contrasts music and life post-storm; one scene plays Rafael Cortijo y su Combo's *Noche de Temporal* ("Night of Tempest," 1970), about the dread of a hurricane night: "Noche de temporal, / que mete miedo a cualquiera. / Allí en Ponce temblaron, / los guapos de la cantera" ("Night of tempest, / that scares anyone. / There in Ponce they trembled, / the tough guys of the quarry").

Córdova highlights inherent disaster inequalities, both in the immediate impact and for long-term consequences. As critic Alessia Vignoli says of Caribbean culture: "catastrophes can be used by the writer to denounce social inequalities. She can also give voice to an individual or collective resistance"; 202, 150, our translation).

The theme of disaster capitalism runs throughout the exhibit. Migration and government inefficiency are recurring topics in Córdova's testimonies. A woman named Maggie says: "Te juro que sí anuncian otro [huracán], yo me voy, en un avión. El gobierno no está haciendo lo correcto. Venden el país al mejor postor" ("I swear if they announce another [hurricane], I'm leaving on a plane. The government is not doing right. It's selling the country to the highest bidder"). Maggie's nameless relative talks about independence culture; including Albizu, remarking that even clothing signals affiliations, specifically, white button-down shirts are a uniform of local bureaucrats; the scene then cuts to men wearing white shirts. Maggie's relative says these were the people who caught Trump's paper towel rolls and bitterly comments on climate change saying he would sell his waterfront property to outsiders because it will be gone in 10 years. He complains about taxes, the Junta and the government, and the entirety of the colonial structure that contributed to the island's vulnerability (Fig. 5.3).

There is much humor in the exhibit. GabriellaTorres-Ferrer's *Untitled (Valora tu mentira americana)* ("Untitled (Value your American Life," 2018), which went viral, wryly mixes the topic of infrastructure (a real-life electrical pole) and politics (an attached poster board lauding U.S. citizenship). The pole itself is a found object, which links to the topic of disaster debris.

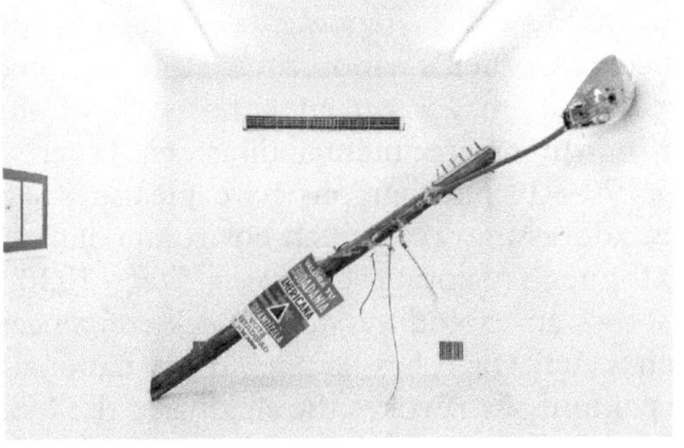

Fig. 5.3 Gabriella Torres-Ferrer, *Untitled (Valora tu mentira americana)*, 2018. Lamppost and corrugated-plastic sign, 116 × 118 × 122 in. (294.6 × 299.7 × 309.9 cm). Whitney Museum of American Art, New York; gift of the Mima and César Reyes Collection, Puerto Rico 2023

Gabriella N. Báez's mixed media *Ojalá nos encontremos en el mar* ("Hopefully, We'll Meet at Sea" (2022) (Guerrero 2022, 70–71) is about family life and migration as red threads connect the photos of the artist and her father, who committed suicide two months before the first anniversary of Maria. As father and daughter appear first on a beach and then in an airport, the metaphorical bloodlines unite themes of family, grief, and migration. The spools of red thread and sewing scissors make this a particularly gendered work, as sewing was traditionally woman's work.

Directly addressing the landscape, Candida Alvarez's *Here to There* (2022) (Guerrero 2022, 66–67) contrasts the traditional idealized landscape versus the post-Maria landscape. Her two-sided painting called *Lomas* with images of a hill in her native town Villalba, presents on one side a pre-hurricane lush land and on the other a post-hurricane denuded view. The real-life impact of the piece's recto/verso technique in semi-transparent fabric allows viewers to gaze through both, merging the landscape before and after the storm.

Despite its title, the Whitney exhibit was also about climate change disaster. Frances Gallardo's paintings *Aerosoles* ("Aerosols") (2021–2022) (Guerrero 2022, 80–81) depicts the Saharan dust clouds that routinely coat the Puerto Rican landscape. Gallardo's "artwork embodies the intersection of political discourse surrounding colonialism and climate change"

(Pérez Martínez 2023, 129). *Aerosoles* uses scientific data, samples of the dust examined at Cornell's nanotechnic lab which become the background for her digital images on an entire wall of the exhibit. Also addressing non-human environmental disasters, Javier Orfón's (2021) (Guerrero 2022, 90–91) photographs have messages written on iconic *cupey* tree leaves, addressing widespread environmental degradation.

Gamaliel Rodríguez's painting *Collapsed Soul* (2020–21) (Guerrero 2022, 96–97) shows an exploding ship, which references colonial shipping and the Jones Act which has hampered the national economy since the 1920s. The painting also evokes the sinking of the USS Maine, which triggered the Spanish-American war that led to Puerto Rico's territorial status (Fig. 5.4).

Tryptic *Fig. 1828 LMM* depicts airport control towers directly referencing the Luis Muñoz Marín international airport, an important symbol of migration. Post-apocalyptic in tone, vegetation sprouts from the

Fig. 5.4 Gamaliel Rodríguez. *Figure 1828 LMM* 2018. Acrylic and ink on paper. Whitney Museum of American Art

towers, signaling that in the long game of planetary history, the non-human will outlive the human. Edra Soto's *GRAFT* (Guerrero 2022, 106–107) was started in 2013 and includes views of hurricane Irma's and Maria's damage seen through viewfinders embedded in wrought-iron and cement fences, melds infrastructure and traditional architecture.

In his installation *Shields/Escudos* (2020) (Guerrero 2022, 87–89) Miguel Luciano makes direct use of the work *La Puerta*—the 2016 art collective Artistas Solidarxs's black and white version of the national flag that mourns the national colonial status. Luciano's semicircle of ten sheet metal plates made from decommissioned school buses, highlights how post-Maria's migration depopulated the archipelago, forcing many school closings.

The theme of grief and remembrance that we see in so much post-Maria art and literature appears literally in Garvin Sierra Vega's *4645* (2019) (Guerrero 2022, 104–105) a digital-image work mourning the number of deaths after Maria, many due to lack of electricity and medical care. The number comes from a study from the Harvard T. H. Chan School of Public Health and the Beth Israel Deaconess Medical Center which calculated there were 4645 excess deaths between September 20 and December 31, 2017. Kishore et al. caution that "this number is likely to be an underestimate because of survivor bias" (2018) since people tended to focus on those who survived rather than on those who did not. Sierra Vega's thirty-seven printed works plus screens display the biting humor in post-Maria social media. They range from caricatures of Donald Trump to clever comments using traditional symbols such as the machete, political party logos, maps—like the main island drowning or Florida pissing on the archipelago—all in contrast to the stark number of the dead: 4645. The prints cover topics like femicides, *perreo* dance protests, and crucially, the topic of community and resilience: "solo el pueblo salva el pueblo" ("only the people save the people").

Elsewhere, Antonio Martorell devoted an entire exhibition memorializing Maria's dead in 2018 with "*¿Quéslaque?*" "*Es* que la..." Martorell specifically denouncing government excuses for the deaths. The pun of the title refers to "esquela" ("obituary") and the artworks consisted of crossed out obituaries. The artist stated the two are interconnected: "la muerte como evidencia de la incapacidad del gobierno de manejar la crisis" ("death as evidence of the government's inability to handle the crisis" (Rodríguez Martinó 2018).

Daniel Lind-Ramos's conceptual lens acknowledges the political, economic, and cultural factors involved in the disaster nation. His exhibit, *El Viejo Griot—Una historia de todos nosotros* ("Old Griot—A Story of All of Us," MoMA PS1 2023) links hurricanes, climate change, and Puerto Rican culture. Lind-Ramos fuses both Catholic (Maria) and Taino (Guabancex) female icons in some of his masterful syncretical large assemblage sculptures [15] which use compositional found objects, debris, and plant materials from the environment.

Lind-Ramos's sculptures are major examples of how the almost incomprehensible vastness of global climate change can be comprehended when grounded in local cultures and history; as Mike Hulme says: "global climate change has to be re-invented as discourses about local weather and about the relationships between weather and local physical objects and cultural practices" (2008, 6). *Baño de María* ("Bain-marie/The Cleansing" 2018–2022; MoMA PS 2023) references the slow warming of the oceans; as Lind-Ramos says, "the reason why those hurricanes appear is because of the warming of the water in the area" (Exhibit audio).

Baño de María is the most Maria-centric and syncretic of the assemblages of course, clearly meshing Catholicism, climate change and hurricanes with its pentagram of spirals or sigmoides echoing the "dynamic interpretation" that Fernando Ortíz noted as universal symbols of hurricanes (1947, 31). Lind-Ramos summarizes his syncretic use of the name: "María, which in the Christian tradition is a character that protects, love, et cetera. On the other hand, María was the name of a destructive force of nature" (Exhibit audio). *Baño de María*, *María de los sustentos* ("Mary of Nourishment" 2021; MoMA PS1 2023), and *María Guabancex* syncretize Taino, Spanish and African cultures. Oddly enough, there are no direct references to the United States, the fourth "story" of the Puerto Rican national house (following José Luis González's metaphor of the nation as a building). This absence is stark in the video *Talegas de la memoria* ("Sacks of Memory" MoMA PS1 2023) which documents part of a performance in Loíza in 2020. The national story begins with a male figure personifying the primeval landscape who calls for the Tainos; the land is later invaded by the Spanish who bring Africans, who at the end of the video take the golden heart symbolic of the nation. Wither the U.S.? Elsewhere in the exhibit, in the *Tocones* video, Lind-Ramos refers to a failed English invasion but again Americans are absent, except as materials in the blue *toldos* which are the Virgin Mary's shawl in *Baño de María*. Lind-Ramos uses toldos strategically in *María de los sustentos* in which they cover shovels representing the manual labor of local communities that cleared debris.

Disaster in the assemblages extends beyond Maria to the colonial dependency it underscored, and also include the Covid-19 pandemic in *Ambulancia* ("Ambulance"; MoMA PS1 2023) and *Alegoría de una obsesión* ("Alegory of an Obssession) (MoMA PS1 2023). The first was fruit of Lind-Ramos's residency in New York just as the Covid-19 outbreak wreaked havoc on that city, and alludes to the continual soundscape of ambulances, i.e., the noise pollution of disaster. The second, *Alegoría*, is humorous, especially the raggedy sneakers at the bottom, stressing that the pandemic affected *gente de a pie* (people on the street).

The hive of cultural activity in the aftermath of Hurricane Maria is a sign of Puerto Rico's cultural resilience, sense of community, and national identity in the face of the displacements caused by disasters and colonialism. The chronic conditions that deepened the crises give immediacy to the work of cultural laborers, as shown by the quick publication of Yarimar Bonilla and Marisol LeBrón's collection of critical essays *Aftershocks of Disaster: Puerto Rico Before and After the Storm* (2019).

Christopher Abram says the unfolding planetary catastrophe "demands the attention of all thinkers and the action of all citizens. Global warming, mass extinctions, overpopulation, and the unsustainable exploitation of the Earth's resources are far too important to be left to scientists or (even worse) politicians" (2019, 20). This is why the urgency of the stories and art examined in this chapter are wake-up calls against ecoerasure about the global climate challenges which will, among other dangers, create supercharged storms in the Atlantic Ocean (Cheng 2022).

The cultural tempest after hurricane Maria exploits disasters as a metanarrative to the story of the Nation. Puerto Rican culture has renewed its engagement with its violent environment, which will increasingly be the bitter fruit of the Capitalocene as oceanic waters warm. Representations of the landscape have changed but, will the garden paradise iallegory ever wholly disappear from the national imagination? Nothing is a monolith and certainly not Puerto Rican culture; for instance, in 2022 it was still possible to see a description of the archipelago as a land, "world famous for its beautiful beaches, its pleasant people and for its benevolent climate that allows you to enjoy its coasts all year round" (García-Moreno Barco, 2022, 145). How long will the forces of touristification and nationalism continue to "forget" the disaster nation?

Notes

1. A disastrous year for the bioregion. Hurricanes Harvey, Irma, and Maria hit the Caribbean and the U.S. coastline, including Texas and the Florida Keys. (Wallace et al.).
2. NOAA's Technical Memoranda shows the estimated pressure at landfall, the height of the coastal waves, the winds on the Saffir/Simpson scale, and measures of inland floods from torrential rains. Governmental and insurance companies' calculations of property damages and losses followed these scientific measurements.
3. https://web.archive.org/web/20250403114741/https://www.elnuevodia.com/entretenimiento/cultura/notas/antonio-martorell-revisa-su-cartografia-en-mapas/.
4. Alejandro Torres-Abreu et al. describe nine exemplary community initiatives: Comité Caborrojeño Pro Salud y Ambiente, Proyecto ENLACE, the Grupo de las Ocho Comunidades (G-8) of Caño Martín Peña (San Juan), Casa Pueblo of Adjuntas, the Corredor del Yaguazo in Ciénaga Las Cucharillas (Cataño), the Huerto, el Vivero y Bosque Urbano de Capetillo (Río Piedras), and the Programa de Educación Comunal de Entrega y Servicios (Humacao).
5. https://web.archive.org/web/20250126151157/https://www.theguardian.com/world/2021/apr/22/hurricane-maria-puerto-rico-trump-delayed-aid.
6. https://web.archive.org/save/https://www.lavanguardia.com/vida/20230315/8827846/puerto-rico-organiza-primera-cumbre-erosion-costera.html.
7. There also is resistance to solar farms. Rooftop distributed solar is what IDEBAJO leaders advocate for. Consult "Visualizing green capitalist renewable energy: Development and grassroots solar community alternatives in Puerto Rico," for more information.
8. Michael Diebert provides details of what transpired with Trump during and after Maria (2019, cf. 116–117).
9. Puerto Rico imports around 80% of its food, by comparison on Sept. 25, 2024 president Abinader said at the UN that the Dominican Republic produces 90.6% of its food.

10. https://web.archive.org/web/20250126151434/https://www.notiuno.com/noticias/economia/dram-tica-disminuci-n-de-poblaci-n-en-municipios-del-sur-y-este-de-la/article_930c0960-0eff-11ec-a1fe-bb92c8c63aad.html.
11. Hunter College's *Centro* resource hub documents the trends on the aging: https://web.archive.org/web/20250126151611/https://centropr.hunter.cuny.edu/projects/disaster-resource-hub/

    In 2023 Puerto Rico has the 10th most aging population in the world. https://web.archive.org/web/20250126151820/https://www.washingtonpost.com/nation/2023/01/13/puerto-rico-hurricanes-climate-elderly/.
12. See Zambrano, Chapter Four "Subversive Interruption."
13. https://web.archive.org/web/20250126152012/http://bpca.ny.gov/places/museums-memorials/hurricane-maria-memorial/.
14. https://web.archive.org/web/20250126152239/https://www.cnn.com/2017/10/16/health/puerto-rico-hospital-ship/index.html.
15. At MoMA PS1 in Queens in 2023.

CHAPTER 6

Conclusion

This book begins with a profetic epigraph from Alejandro Tapia y Rivera, the foundational Puerto Rican nineteenth-century historian, essayist, short story, and novel writer:

> De la tormenta bravía
> del escollo misterioso,
> líbrame, cielo amoroso
> ("From the fierce storm
> from the mysterious shoal,
> deliver me, loving heaven")
> "La plegaria de una virgen" ("A Virgin's Prayer") (*Obras* "Works" 1968; 567)

Tapia includes it in his (for us) aptly titled collection *Adiós al buen tiempo* ("So long to good times/weather," 1856). It illustrates many of the themes that *Disaster Nation* explores throughout five plus centuries of Puerto Rican culture. Firstly, his *plegaria*[1] reflects the thinking of his times regarding disasters; for instance, the quintessentially faith-based attitude for dealing with the climate anxiety of natural disasters which were unknowable and mysterious. We now, in the full Capitalocene, regard these disasters as far from "natural" due to the depredations our species has visited upon our planet. In the case of Puerto Rico, as Tapia would

M. Acosta Cruz, *An Ecocritical Study of Puerto Rican Culture*,
https://doi.org/10.1007/978-3-032-04044-2_6

agree, colonialism is a mega-disaster on top of hurricanes and earthquakes. We have seen time and again in Puerto Rico's creative efforts the aching need to survive the calamity that is colonialism.

Tapia's *plegaria* is a synecdoche for how earlier cultural responses to hurricanes asked for heavenly protection from these catastrophes, while at the same time viewing the causes as supernatural, regarded as punishments for society's sins. Have people stopped praying even though now we face hurricanes armed with modern technologies and scientific knowledge that allows us to see the monsters approach days ahead of time? We suspect the answer is no.

Storms are nowadays no less fierce than Tapia says, and in fact are getting worse, as the scientific studies we quoted throughout this book prove. If we take Tapia's word "shoal" to be an area mostly underwater, a shallow place that is a danger to navigation, it can equally apply to our twenty-first century consternation and alarm because of the shifting oceanic environments of the Capitalocene. Tapia was focused on Puerto Rico, but all our planet is in danger from the not-so metaphorical shoals of rising waters.

Finally, the title of Tapia's prayer, "La plegaria de una virgen" grounds us in one of this book's main themes, the patriarchal feminization of the landscape, which puts a burden of representation on women, often to the detriment of real women's lives. The "isla doncella" trope of the virginal woman that is also somehow tied to the landscape is traced in this book from the earliest colonial writing to present day—as the alternative national anthem "Verde luz" shows. Much as we love them, Luis Palés Matos's wild hurricane women also reinforce stereotypes of women as elemental parts of an often inscrutable and violent Nature. In contradistinction to masculinist Androcene tales, we saw how powerful women writers reclaim Nature as an agentic force that allows them to express women's and the Nation's experiences of the environment, as Julia de Burgos's aquatic ecopoetics show. The triumvirate of Rosario Ferré, Magali García Ramis, and Ana Lydia Vega use the lenses of the Androcene and material feminism to explore the environment and grant agency to forces beyond simply the human.

The explosion of art after hurricane Maria in 2017 was the motivating force behind this book. It led this author to wonder why ecoerasure, i.e., the cultural forgetting of the violent environment, happens in so much of Puerto Rican culture. We have sampled much of the culture during the nation-building period of the eighteenth and nineteenth centuries, as we payed attention to feelings about the *patria* that are still vital, because love of Land continues to be as powerful a force as religion. The post-Maria period propels our discussions to newer disaster themes that uplift the Nation with the kinds of hope needed to help a disaster nation: community reconstruction, climate and energy justice, and decolonizing resilience.

This does not mean old themes are forgotten, since newer culture works are part of a continuum of how the nation has long grappled with its environment, often with ecoerasure. Lest we forget, devastating earthquakes are very much still an unknowable part of Puerto Rico's ecological geology and inspire the same climate anxiety we have seen since at least Fray Iñigo y Abbad. Even more vitally, the complicated interrelations between so-called natural disasters and colonialism are amply documented throughout *Disaster Nation*.

With our term ecoerasure, we clarify the many cultural processes that enable the "forgetting" that Ernst Renan cautioned is necessary for national formation. We have seen that Puerto Rico's national culture is a complex web of overt and hidden references to the metanarrative of the environmental disasters of the Capitalocene. This book has explored stories, symbols and images that show how and why Puerto Rican culture has both elided and highlighted disasters, which are, paradoxically, key to the nation's formation and to its people's resilience. By looking at the meteorological and historical data we traced how Puerto Rican creators often fall back on ecoerasure despite how hurricanes and earthquakes at times make of the Beloved Land a hell, instead of a paradise.

Our critical scaffolding has relied on ecocritical tools in order to study these five plus centuries of history, literature, and art that make Puerto Rico the poster child of ecoerasure. These critical lenses are interdisciplinarity, with particular explorations of paleotempestology, and meteorological data and history; ecofeminism—particularly material feminism; close readings; and postcolonial scrutiny centering on environmental justice, especially Boricua theories of climate and energy justice in the twentieth and twenty-first centuries, which frame our analysis throughout. We consider that so-called natural disasters are made worse by the most negative aspects of the nation's complicated colonial situation.

Since the twentieth century creators have been willing to confront the structural and systemic causes of disaster, in other words, how colonialism and climate change are interwoven. As we have seen, literature and historical writing take circular routes to disaster: early descriptions of hurricanes and earthquakes fully highlight the dramatic shock and awe of hurricanes to remind the Spanish colonial authorities of the precarious living in the colony. The priorities of nation-building and of Love of Land led in the nineteenth century to enduring depictions of a fictionally placid Nature despite ecoreality, as our Appendix X proves. Puerto Rican historical writing could be peppered throughout with the phrase "and then a hurricane happened." We find 2017 to be a signal year when many works returned to early colonial themes with post-Maria descriptions of the trauma of the disaster nation and of its people's resilience.

Meanwhile, art took a more linear route, from very few portrayals of disaster (Campeche stands out) to nineteenth-century placid depictions of Nature required for national formation, to post-Maria's proliferation of representations of the paths of calamity. Be it literature, or history, or art, we have become reacquainted with hurricanes as key characters in the Nation, from Tufiño's *La plena* monster to the unavoidable name that is "Maria" after 2017.

Can fictions and images of a garden of eternal springtime survive persistent colonialism, superstorms, Saharan dust, polluted oceans, and unpredictable earthquakes all in a colony whose infrastructure depends on a fossil fuel-based economy? The environmental humanities give us new urgency to confront these challenges by reading Puerto Rico as a case study on Capitalocene environmental depredations in the Caribbean. Literature and the arts should now produce, in Puerto Rico and everywhere, gripping stories and impactful images to tie our inherent interconnected vulnerability to global phenomena, paying special attention to that which can be mitigated. Ecoerasure, as detrimental as it is for preparedness and resilience, should not persist in a nation that is punished by both its bioregion and the exploitation of U.S. colonial oversight. An ecocritical eye will lead us to stories, symbols and images that will inform us of the peril of living in disaster nations.

Notes

1. The *plegaria* is from a book that mixes history and other kinds of writing, so it may have been a popular prayer he compiled, though he does not label it as such.

Appendix

Catalogue of Disasters Alongside Major Works of Puerto Rican Culture

This catalogue shows the history of disasters in Puerto Rico so readers can assess how an ecocritical context is represented, or not, in the cultural products they study. That is why this Appendix offer a compilation of scientific and historical disaster information. We interweave historical meteorology with significant works of history, literature and art for readers interested in environmental approaches to culture. We also document infrastructure and technology that affect disaster preparedness, recovery, and resilience. We regard colonialism throughout as a mega-disaster.

The disaster nation is well-documented; climate historians Michael Chenoweth and Ian Howard state that: "The central and eastern Caribbean Sea region has the longest continuous record of hurricane impacts anywhere in the world, extending back to 1494" (2023, 1). Meteorological and historical sources show the long history of disasters in Puerto Rico, catalogued here in chronological order; with selected descriptions of all recorded hurricanes, earthquakes (and epidemics). NOAA's administration provided crucial support for this project. Michael Chenoweth generously shared details on much hurricane data and history. We rely on NOAA's database of Historical Hurricanes and Tropical Systems from 1842 onward.[1] Hurricane information before 1842,

M. Acosta Cruz, *An Ecocritical Study of Puerto Rican Culture*,
https://doi.org/10.1007/978-3-032-04044-2

however, comes from multiple sources which are sometimes contradictory (for example, the year 1751). Epidemics are also disasters of the nation often due to bad urban planning and are not "natural" in the way of hurricanes and earthquakes—although government mismanagement and colonialism are major factors in cyclonic and geologic cataclysms.

This catalogue underlines an important thesis, culture tends to ecoerase hurricanes when there were significant gaps between hurricanes. Michael Chenoweth and Ian Howard explain that: "The frequency of hurricane landfalls and impacts in the CECR [Central & Eastern Caribbean Region] was lower in the twentieth century than any time since 1494 A.D" (2023). These two climate scholars trace the historical gaps in hurricanes from 1514 to 1989 (Chenoweth and Howard 2023, 5).[2] Significant gaps between hurricanes frequently coincide with ecoerasures in Puerto Rican culture.

The Saffir-Simpson Scale has been in use since the 1970s:

Category 1 = 74–95 mph
Category 2 = 96–110 mph
Category 3 = 111–130 mph
Category 4 = 131–155 mph
Category 5 = > 155 mph
(Blake 2011)

A major hurricane is Category 3, 4, or 5. There is discussion of adding a Category 6. Storm winds kill people and destroy habitats and ecosystems, but there are also deadly storm surges, waves, floods, mud slides, downed power lines, extended lack of electricity, and other traumatic outcomes.

Although seismic activity is monitored by the Puerto Rico Seismic Network of UPR-Mayagüez as well as the U.S. Geological Survey's National Seismic Hazard Maps, earthquake prediction remains challenging.[3]

Works of history, literature, and art are labeled according to genre or media.

1493 (correspondence) — Christopher Columbus, first letter to Luis Santángel.

1508, Aug. 26, hurricane San Roque I — This is the first hurricane to strike after the Spanish arrived. Millás (1968, 44–45). Salivia states this hurricane "passed through the south of the island... when don Juan Ponce de León began his immortal work of colonization" (1950, 1). Vivas dates it at Aug. 16 (1962, 319).

1508 (correspondence) — Juan Ponce de León.

1510, Smallpox epidemic — Abbad reports it (1788 90).

1514, Unnamed hurricane — Salivia quotes from a letter to the King from Royal Officials: "many Indians died, due to illness caused by the storm, as well as the lack of food" (1950, 10). Caldera Ortíz quotes the same source (2014, 16), as does Middledyk (1903, 42). Millás (1968, 48). Miner Solá says there is some doubt about the year in which it occurred (1995, 13).

1515 — "Two hurricanes... put survival in jeopardy, destroying agricultural production and killing numerous indigenous people" (Burset 2022, 46). Vivas calls it San Laureano dating it at July 4, 1515 (1962, 319).

1526, Oct. 14–15, hurricane San Francisco — Asenjo dates it as 1527: "A strong hurricane that lasted 24 hours and destroyed the crops. Most of the city and the church were blown down" (1886, 96). Miner Solá (1995, 13). Tapia quotes Juan de Vadillo's letter to the emperor, dated March 20, 1527:

> On the night of October 4 last, a tempest of wind and water, here called a hurricane, which lasting 24 hours, 'destroyed the greater part of this city, including the church, and did so much damage to the estates in the country on account of the great currents in all the rivers, that nothing similar can be remembered in this island'; many rich people have grown impoverished (1854, 295).

1530, Three hurricanes — Sources report between three and five storms that year. Caldera Ortíz says accounts of the triple disaster of 1530 are "immense and at the same time extensive" (2014, 15). Middledyk writes: "The storms have destroyed all the plantations, drowned many cattle, and caused a great dearth of food.... Everybody has been impoverished and is thinking of leaving. There are no more Indians and the land must be cultivated with negroes.... The city begs that the payment of all debts be postponed for three years" (1903, 62). Miner Solá says there were three hurricanes (1995, 13). Ubeda y Delgado also says three storms left the island almost ruined (1878, 14).

1530, July 26, hurricane Santa Ana — Chenoweth dates it July 28 (email); storms were and still are rare in summer. Destroyed half the capital houses (Miner Solá 1995, 13).

1530, Aug. 5, hurricane San Hipólito — Neely says Puerto Rico was left "hungry, defenseless, and subject to attacks from the Caribs of the lesser Antilles, whose own economy had also been disrupted by the storms" (2017, 178). Tapia says then-governor Lando wrote a letter to the empress of Spain in 1533 describing the poor state of the colony, linking storms to infrastructure debt: "Much of the cause has been the storms of past years which destroyed the farms and they had to incur maintenance [for buildings], the debts grew" (1854, 299). Millás labels it "a moderate intensity hurricane, with much rain" (1968, 303). Vivas dates it at Aug. 22 (1962, 319).

1530, Aug. 31 or Sept. 1, hurricane San Ramón — Millás calls it "a very weak cyclone, with much more rain than violent winds" (1968, 60). Miner Solá says it damaged crops and cattle (1995, 14). Middledyk quotes a contemporary source:

> [crown officers wrote in 1534] the island is becoming depopulated, gold is diminishing, the Indians are gone.... The people, ruined by the hurricanes of the year 1530, thinking that they might find other gold deposits, bought negroes on credit at very high prices to search for them.... Some are fugitive in the mountains, others in prison, others again have stolen vessels belonging to the Administration and have gone with their negroes who knows where. With all this and the news from Peru, not a soul would remain if they were not prevented from leaving" (1903, 73).

1530, Sept. 10, Unnamed hurricane — Acosta y Calbo lists hurricanes as one of the main causes of "ruin and backwardness" of the period between 1521 and 1530 (1866, 118). Middledyk reports dire conditions in the colony: "they met many adverse conditions: incursions by the Caribs, hurricanes in 1530 and 1537, emigration to Peru and Mexico, internal dissensions, and last, but not least, heavy taxes" (1903, 162–163).

1535, Sept. 22, hurricane San Leoncio — Millás finds information inconclusive (1968, 64). Vivas dates it Sept. 12 (1962, 319).

1535 (history) — Gonzalo Fernández de Oviedo y Valdés, *Historia general y natural de las Indias, islas y tierra-firme del mar océano. Primera parte* ("General and Natural History of the Indies, Islands and Mainland of the Ocean Sea. Part One").

1537, between June and Sept., Unnamed hurricanes — Millás finds information about three hurricanes uncertain concluding it was instead "heavy rains" (1968, 65–66). Chenoweth believes it was not a hurricane (email). Tapia notes there were three storms (1854, 314). Miner Solá reports many slave deaths (1995, 14) and Acosta y Calbo says many of them drowned (1866, 433). Middledyk quotes sources that say three hurricanes in two months devastated the region: "The floods swept all the plantations along river banks, many slaves and cattle have drowned, destitution and poverty are universal. Those who wanted to leave the island before are now more than ever anxious to do so" (1903, 62). Vivas dates hurricane San Pío to July (1962, 319).

1545, Sept. 17–18, Unnamed hurricane — "it was not a severe storm, rather one of great diameter and rather slow movement" (Millás 1968, 70).

NOTE: Between 1545 and 1568 there is a gap in well-documented disasters.

1527–1561 (history) — Fray Bartolomé de Las Casas. *Historia general de las Indias* ("General History of the Indies").

1568, Sept. 3, hurricane San Bartolomé — First hurricane named after a Catholic saint; earlier hurricane names assigned retroactively (Miner Solá 1995, 14). Millás (1968, 84); Tapia (1854, 458); Vivas (1962, 319).

1575, Oct. 1, hurricane San Mateo — Middledyk (1903, 218); Millás (1968, 86); Vivas (1962, 319).

1582 (history) — Juan Ponce de León Troche and Antonio de Santa Clara, *Memoria de Melgarejo.* ("Melgarejo's Memoir").

1589 (history) — Juan de Castellanos. *Elegías de varones ilustres de Indias* ("Elegies of Illustrious Men of the Indies").

1591, Sept. 21, Unnamed hurricane — Chenoweth (email); Medrano Herrero (2008, 30).

1601–1615 (history) — Antonio de Herrera, *Historia general de los hechos de los Castellanos en las islas, i tierra-firme del Mar Océano* ("General History of the Deeds of the Castilians in the Islands and Mainland of the Ocean Sea").

1615, Sept. 7 or 8, Unnamed hurricane; Earthquake — Acosta y Calbo mentions a "rigorous storm" (1866, 433). Medrano Herrero quotes a letter mentioning an earthquake (2008, 21, 23).

1615, Sept. 12, hurricane San Leoncio II — Middledyk: "It did so much damage to the cathedral that it was necessary partly to cover it with straw and write to his Majesty asking for a donation to repair it. With his accustomed generosity, he gave 4,000 ducats (1903, 218). Medrano Herrero (2008, 20); Miner Solá (1995, 15) confirm. Millás calls it normal or severe (1968, 100). Vivas (1962, 319).

1625 (history) — Larrasa, Diego de, *Relación de la entrada y cerco del enemigo Boudoyno Henrico, general de la armada del príncipe de Orange en la ciudad de Puerto Rico de las Indias* ("Account of the Entry and Siege by the Enemy Boudoyno Henrico, General of the Army of the Prince of Orange in the City of Puerto Rico of the Indies" (Tapia 1854, 416–433).

1626, Sept. 15, hurricane San Nicómedes — Around 30 deaths; 30 ships sank in San Juan Bay. Many houses destroyed and the belfry and bells of a convent fell (Millás 1968, 106). Santo Domingo convent's bell tower fell on the chapel roof; floods caused crop damages; 38 people killed (Salivia 1950, 44). Vivas calls it Santa Catalina (1962, 319).

1641 Unnamed hurricane — "destroyed the cocoa crop" (Brau 1917, 137).

1642, Sept. 14–15, Unnamed hurricane — Medrano Herrero cites damage to a church (2008, 28). Millás calls it a "great storm which destroyed a church" and wreaked "havoc"; pigs were particularly affected (1968, 114–115). Cacao crops severely damaged (Miner Solá 1995, 15). Bishop Fray Damian Lopez de Haro's letter states "the greatest work (of destruction), was that of the storm and tempest which occurred in the year 42... it destroyed the transept, pulled out many trees and huts and did so great a havoc that it left the soil sterilized (barren) until today... all the pigs that were abundant in this island, with the tempest of the year 42 the greater number died, and (those that remained) fled to the woods close by" (Tapia 1854, 445). Vivas (1962, 319).

1644 (correspondence) — López de Haro, Fray Damián. Letter (Tapia 1854, 445).

1647 (correspondence) — Diego de Torres Vargas. *Descripción de la isla y ciudad de Puerto Rico, y de su vecindad y poblaciones, presidio, gobernadores y obispos; frutos y minerales* ("Description of the Island and City of Puerto Rico, and of its Environs and Towns, Fortress, Governors and Bishops; Fruits and Minerals") (Tapia 1854, 447).

1657, Aug., Unnamed hurricane — Millás calls it "normal" (1968, 126). Miner Solá says damage to crops caused hunger for years (1995,

15). "[A] storm all over the island. Most of the cocoa and other fruit trees were blown down" (Salivia 1950, 50).

1665, Unnamed hurricane — "[T]he French Governor of Tortuga, Beltrán Ogeron, planned the conquest of Puerto Rico. He appeared off the coast with 3 ships, but one of the hurricanes so frequent in these latitudes came to the island's rescue. The ships were stranded, and the surviving Frenchmen made prisoners. Among them was Ogeron himself" (Middledyk 1903, 103).

1673, Unnamed hurricane — Chenoweth (email). Millás narrates much the same tale from 1665: "Monsieur Ogeron... was abruptly surprised with a violent storm. This increased to such a degree that it caused his new frigate to strike the rocks in neighboring islands, called Guadanillas, where the vessel broke into a thousand pieces. Yet being near the land of Puerto Rico, all his men escaped in boats" (1968, 138–139).

1678, Unnamed hurricane — Middledyk tells a similar tale:

> Duke Estren, an English commander, with a fleet of 22 ships and a body of landing troops appeared before San Juan and demanded its surrender, but that, before the English had time to land, a violent hurricane occurred which stranded every one of the British ships on Bird Island. Most of the people on board perished, and the few whose lives were saved were made prisoners of war (1903, 218).

1679 (correspondence) — Anonymous letter of complaint against the colonial regimes of don Gaspar de Arteaga (1670–1674) and don Juan de Robles Lorenzana (1678–1683) (Rivera de Álvarez 37, note 12).

1685, around Sept. 10, Unnamed hurricane — Chenoweth (email).

1690 (history/fiction) — *Infortunios que Alonso Ramírez natural de la ciudad de San Juan de Puerto Rico padeció* ("The Misfortunes of Alonso Ramírez Natural of the City of San Juan de Puerto Rico"). Attributed to Carlos de Sigüenza y Góngora.[4]

1692, Sept. 13, Unnamed hurricane — Caldera Ortíz concludes the amount of documented hurricane activity during the seventeenth century was "light" (2014, 49). Chenoweth (email).

1713, Sept. 5–6, hurricane San Zacarias — Chenoweth says it is "reputed to be one of the worst to that time" (email). Millás (1968, 168–169). Caused a cyclonic tide (Miner Solá 1995, 16).

1713, Oct. 9–10, hurricane San Cándido — Millás (1968, 15).

1718, Sept. 7, hurricane Santa Regina — Salivia says it, "flooded the chapel at the convent.... the convent was unroofed" (1950, 57). Millás (1968, 15), Vivas (1962, 319).

1720, Oct. 9–10 Unnamed hurricane — Sea disaster, a man-of-war sank with all on board off P.R., however, Millás says the storm is not mentioned in hurricane lists (1968, 176–177). Boose.

1728, Aug. 30, Unnamed hurricane — Millás (1968, 181–182).

1729, Aug. 16, Unnamed hurricane — Chenoweth (email).

1730, Aug. 29–30, Unnamed hurricane — Millás (1968, 182).

1738, Aug. 30–31, hurricane Santa Rosa — This storm ended a drought in the southern part of P.R. for years (Caldera Ortíz 2014, 51). Millás says it was normal size but "severe" (1968, 196). There were "important damages" in San Juan (Miner Solá 1995, 16).

1738, Sept. 12, hurricane San Leoncio III. Plague — The colony's economy already was in bad shape, after the storm "a plague of worms created by the excess of humidity in the terrain" ate planted seeds (Caldera Ortíz 2014, 52). Ponce and Coamo abandoned (Millás 1968, 197).

1740, Aug. 3, hurricane San Esteban — Ponce suffered damages (Millás 199).

1740, Sept. 11–12, hurricane San Vicente — Chenoweth states it was, "said to be worse than the one in 1738" (email). Middledyk says the precise date is unknown, and "this hurricane destroyed a coco-palm grove of 5 or 6 leagues in extent near Ponce" (1903, 218). Millás (1968, 17). Salivia notes it "destroyed all the fruits and killed most of the cattle" (1950, 69).

1742, Aug. 27–28, hurricane San Judas Tadeo — "[A]n English ship sank in front of the coast of Coamo and Ponce." (Brau, 1917, 173). Millás (1968, 202), Vivas (1962, 319).

1751, Aug. 18, hurricane San Agapito. Earthquake — Asenjo says "A hurricane struck all over the island" (1886, 87–88). Chenoweth says it was an earthquake, not a hurricane (email). "The ship "Gran Canal" sank in the Bay.... The crew was saved, except for the captain, thanks to the efforts of a sailor called Victor Rojas" (Coll y Toste 1916, 170). Millás calls it a weak storm (1968, 206). "[T]he San Agapito hurricane occurred along with an earthquake. It was felt on all the Antilles, particularly on Jamaica and Santo Domingo" (Salivia 1950, 75). Earthquake reported as this storm passed (Miner Solá 1995, 17).

1753, Oct. 2–3, Unnamed hurricane — Chenoweth (email).

1765 (history) — Alejandro O'Reilly. *Relación circunstanciada del actual estado de la población, frutos y proporciones para fomento que tiene la Isla de San Juan de Puerto Rico.* ("Detailed Account of the Current State of the Population, Fruits and Proportions for Development in the Island of San Juan of Puerto Rico").

1766, Sept. 19, hurricane San Genaro — Asenjo calls it a strong hurricane that destroyed Cangrejos (in Santurce), Río Piedras, Loíza, Fajardo, Caguas, Las Piedras, and Guayama (1886, 89). Struck the eastern part (Millás 1968, 222). Crop damages and blocked roads (Miner Solá 1995, 18). "[D]efinitely one of the worst hurricanes ever experienced in the island. After the storm, there was a worm pest that destroyed all the seeds" (Salivia 1950, 77).

1766, Oct. 7, hurricane San Marcos — Traversed the region (Millás 1968, 224). Vivas (1962, 319).

1767, Aug. 7, hurricane San Cayetano — "A strong hurricane that destroyed the plantain trees. Many heads of cattle were lost" (Asenjo 1886, 86). Millás believes it had strong gales (1968, 224). Neely says it passed to the south causing flooding and crop damage (2017, 200).

1768, Oct. 8–9, hurricane San Marcos — Millás (1968, 225).

1772, July 16 Unnamed hurricane — Passed over the north seas (Millás 1968, 230). Vivas call it Del Carmen (1962, 319).

1772, Aug. 28 or 29, hurricane San Agustín — Abbad witnessed it (431). "In 1774, Fray Manuel Jimenez Perez informed the king of the damage that many churches received" (Coll y Toste 1916, 347). Millás dates it Aug. 28-Sept. 3 (1968, 235). Vivas (1962, 319).

1772, Aug. 31-Sept. 1, hurricane San Ramón — Garcia. Miner Solá calls it a lesser storm (1995, 18). Vivas (1962, 319).

1775, Aug. 1, hurricane San Pedro — Boose says to the northeast: "destroyed many countryside houses and flooded many places. A lightning bolt killed a prisoner and injured nine other prisoners." Floods killed many cattle and smaller animals (8/20/1775). Millás, however, says it passed to the south (1968, 237). Garcia; Vivas (1962, 319).

1776, Sept. 7 hurricane Santa Regina — Destroyed casabe, bananas and rice crops (López Cantos 2000, 218). Vivas (1962, 319).

1780, June 13, hurricane San Antonio — "Strong hurricane that destroyed all the countryside of the island" (Asenjo 1886, 76). An "early date" Millás (1968, 249). It "caused deaths and losses" (Neely 2017, 207). Vivas calls it San Paulino (1962, 319). Garcia, Millás (1968, 249).

1780, Oct.14, hurricane San Calixto. Earthquake — Middledyk:

> In 1780 a fleet was equipped with the object of definitely destroying all the pirates' nests. The greater part of the garrison, all the Puerto Ricans most distinguished for bravery, intelligence, and experience, took part in the expedition. The fleet was accompanied by the Spanish battle-ship Carlos V, which carried 50 cannon and 500 men. Of this expedition not a soul returned. It was totally destroyed by a hurricane, and the island was once more plunged in mourning, ruin, and poverty, from which it did not emerge till nearly a century later (1903, 134-135).

He calls it "the most terrible cyclone of modern times" noting it "seems to have embodied all the horrible scenes that attend a phenomenon of this kind" (1903, 289). "The Great Hurricane of the Antilles" (Millás 1968, 253). Sheets and Williams say that this hurricane devastated the Caribbean and parts of the U.S., causing 22,000 deaths in total, and passing near Puerto Rico. Also, that it was a slow storm (2001, 18) and deaths were probably underestimated since many of the ships lost carried slaves, who were not always counted (2001, 20). Neely says it caused deaths and losses in Puerto Rico.[5] Miner Solá says it came with an earthquake (1995, 18).

1785, Aug. 15, hurricane San Lupo — Boose says Loiza, Humacao, and Guayama plantain crops were damaged. He also notes damage "at the San Geronimo fort, located at the left side of the Boquerón of San Juan.... The whole structure was damaged, moved from its original position and needed to be reconstructed" (9/5/1785). Shipwrecks in Humacao and Guayama: "Inside this last one, 12 dead bodies were found as well as a cargo of flour, sugar and rum" (Boose, 9/5/1785). Vivas (1962, 319).

c. 1785 (painting) — José Campeche: *Dama a caballo* ("Lady on Horseback")

1785, Sept. 25–26, hurricane San Lupo [II] — Miner Solá says it caused severe damage (1995, 18). Garcia, Millás (1968, 270) and Neely (2017, 215).

1787, Aug. 24–25, Unnamed hurricane. Earthquake — The earthquake caused the collapse of the parish of Arecibo and damaged that of Ponce, it affected San Juan's defensive system, according to the repair expenses.[6]

1787 (painting) — José Campeche: *San Emigdio de Áscoli* ("Saint John Emigdio of Ascoli").

1788, Aug. 16, hurricane San Roque — Vivas (1962, 319).

1788 (history) — Antonio Valladares de Sotomayor. *Historia Geográfica, Civil y Política de la Isla de San Juan Bautista de Puerto Rico* ("Geographical, Civil and Political History of the Island of San Juan Bautista of Puerto Rico").

1788 (history) — Abbad y Lasierra, Fray Iñigo. *Historia geográfica, civil y natural de la isla de San Juan Bautista de Puerto Rico* ("Geographical, Civil and Natural History of the Island of San Juan Bautista Puerto Rico"). Published in 1866.

c. 1790 (painting) — José Campeche: *El Salvamento de Don Ramón Power* ("The Rescue of Don Ramón Power y Giralt").

1791 Oct. 25, Unnamed hurricane — Chenoweth (email).

c. 1792 (painting) — José Campeche: *El Gobernador Don Miguel Antonio de Ustáriz* ("Governor Don Miguel Antonio de Ustáriz.").

c. 1792 (painting) — José Campeche: *Retrato de dama* ("Portrait of a Lady").

c. 1795 (painting) — José Campeche: *Don José Mas Ferrer* ("Don José Mas Ferrer").

c. 1800 (painting) — José Campeche: *Gobernador Don Ramón de Castro* ("Governor Don Ramón de Castro").

1804 Sept. 4–5, hurricane Santa Rosalía — "A great cyclone" (Middledyk 1903, 219). Garcia; Vivas (1962, 320).

1804, Sept. 21 hurricane San Mateo — Vivas (1962, 320).

1805, Sept. 11, hurricane San Vicente — "The south coast experienced one of the most powerful hurricanes in its history" (Asenjo 1886, 92). "Sugar mills severely affected" (Salivia 1950, 115).

1806, Sept. 12, hurricane San Vicente [II] — Passed to the southwest (Chenoweth email). Great damage to Ponce (Miner Solá 1995, 20). Salivia says it destroyed Ponce (1950, 116).

1807, Aug. 18, hurricane San Jacinto — "In Aguadilla, a river was formed in the middle of the town" (Asenjo 1886, 87–88). In Aguadilla: "The river destroyed the rice and cane crops. Many heads of cattle were killed. Many landslides destroyed crops that were untouched by the hurricane. Many houses were dragged by the river" (Boose, 8/26/1807). Coll y Toste says it lasted 50 hours: "more than 100 houses were blown down" (1916, 43). "[L]oss of life was big." (Garcia). Lasted 50 hours causing great loss of life and crop destruction (Miner Solá 1995, 20). Vivas calls it San Agapito (1962, 320).

1809, Aug. 2, Unnamed hurricane — "Great death toll" in Puerto Rico (Blake 2011).

1809, Sept. 2, hurricane San Esteban — Vivas (1962, 320).

1812, Unnamed hurricane — Did not affect the whole region but caused crop damage (Asenjo 1886, 88). Asenjo says it was two hurricanes, July 23 and August 12; the first to the south, the other to the west (1886, 90); Vivas concurs (1962, 320). Salivia dates it July 23, 1812, and says it passed to the south (1950, 126).

1813 and 1814 hurricanes — Middledyk observes the coincidence that "Both hurricanes happened on the same date, that is, the 23d of July. Yauco and San German suffered most" (1903, 219).

1816, Sept. 16, hurricane San José de Cupertino — Boose, Garcia.

1819, Sept. 22–23, hurricane San Mateo — "more severe to the north" (Chenoweth email)). García: "destroyed most of the houses and crops." Middledyk says it caused "extraordinary damages on the plantations" (1903, 218). Miner Solá calls it severe (1995, 22).

1824, Sept. 9, hurricane — Boose says most haciendas suffered "severe damage" (9/11/1824). Pirate drama: "passing through Cabo Rojo it sank the ship of the pirate Cofresí" (Miner Solá 1995, 22). Vivas labels it La Monserrate (1962, 320). Garcia.

1825, July 26–27, hurricane Santa Ana — Caldera Ortíz says official reports routinely exaggerated disaster to draw the monarchy's attention. He lists 323 deaths; 347 wounded in Yabucoa alone (2014, 76). He says that, "The psychological impact caused by this storm was a lasting one and the oral story remained for many decades" (2014, 77). According to him it was very potent since the governor reported more than 1,500 palm trees uprooted and these trees are, "well-known for the resistance to the winds" (75). Middledyk says it destroyed "Patillas, Maunabo, Yabucoa, Humacao, Gurabo, and Caguas.... The royal *Fortaleza* (the present Executive Mansion) suffered much, also the house of Ponce" (1903, 219–220). Brau calls it a violent cyclone (1917, 241). García estimates it as Category 4.

1827, Aug. 17–18, hurricane San Jacinto — Boose (8/25/1827). Garcia.

1829 Unnamed hurricane — Losses in an *hacienda* (Boose 8/11/1829).

1832 (poem) — María Bibiana Benítez. "La ninfa de Puerto Rico a la justicia" ("The Nymph of Puerto Rico to Justice").

1835, Aug. 13, hurricane San Hipólito. Epidemic — "Caused the loss of almost all agriculture. There is a scarcity of food" (Coll y Toste 1916, vol. 7, 122). Caldera Ortíz (2014, 81) and Miner Solá (1995, 23) note an epidemic due to animal carcasses in Vega Baja.

1837, Aug. 2–3, hurricane Nuestra Señora de los Ángeles — Technology: Caldera Ortíz says it was a memorable storm both because of its "peculiar" name and because of the use of the barometer; adding it must have been Category 2 (2014, 82). Middledyk says, "All the ships in the harbor of San Juan were lost" (220). Miner Solá confirms the barometer was first used (1995, 23). Acosta y Calbo (434); Vivas (1962, 320).

NOTE: Between 1837 and 1846 there is a gap in well-documented disasters.

1843, November (anthology) — Francisco Pastrana, et al. *Aguinaldo puertorriqueño* ("Puerto Rican Christmas Song/Gift").

1844 (anthology) — Manuel A. Alonso, et al. *Album puertorriqueño* ("Puerto Rican Album").

1845 (anthology) — Manuel A. Alonso. *El Gíbaro* ("The Gíbaro").

1846, Sept. 12, Hurricane San Leoncio IV — Vivas calls it San Vicente (1962, 320). Salivia (1950, 179).

1846 (anthology) — new *Aguinaldo puertorriqueño.*

1846 (anthology) — *Cancionero de Borinquen* ("Songbook of Borinquen").

1848, Aug. 22–23, Unnamed hurricane — Chenoweth (email).

1850, Nov 25–30, Unnamed hurricane — Chenoweth (email).

1851, Aug. 18, hurricane San Agapito — "caused considerable damage" (Middledyk (1903, 220). Garcia; Vivas (1962, 320).

1851, Earthquakes — "in 1851 there were 6 earth tremblors" (Salivia 1950, 181).

1852, Sept. 5, Yellow fever epidemic. Hurricane San Lorenzo — NOAA databases start with this cyclone; category 1, 800 deaths (Blake 2011). Vivas places it two weeks later and labels it San Mauricio (1962, 320). "As a result of the storms of 1852 we had another misfortune, the yellow fever epidemic" (Salivia 1950, 188).

1852 (novel) — Alejandro Tapia y Rivera. *La Palma del Cacique* ("The Palm of the Taíno Chieftain").

1853, Oct. 26, hurricane San Evaristo — Vivas (1962, 320).

1854 (history) — Alejandro Tapia y Rivera, et al. *Biblioteca histórica de Puerto Rico* ("Historical Library of Puerto Rico").

1855, Cholera, epidemic — "[H]orrible pandemic of morbid cholera caused more than 30,000 victims" (Salivia 1950, 188). Infrastructure:

Picó blames the cholera outbreak on the Spanish government's unhygienic city planning that allowed shantytowns to proliferate (1990, 181).

NOTE: Between 1855 and 1867 there is a gap in well-documented disasters.

1858 — Technology/Infrastructure: Morse installs first telegraph lines.

1863 (novel) — Eugenio María de Hostos. *La Peregrinación de Bayoán* ("The Pilgrimage of Bayoán").

1867, Oct. 29, hurricane San Narciso, Earthquakes — Category 3 (Blake 2011). Brau (1917, 264), Miner Solá (1995, 24), and (Salivia 1950, 1950, 199) report an earthquake followed. Salivia calls the storm "abnormal, because of its excessive violence with such an insignificant meteoric causality" (1950, 188). He quotes a source of the era that there were: "dead: 211. injured 17. Bruised 31. Damage to public buildings: 16 churches, 13 houses out of six, five butcher shops and eight cemeteries. In ports: 14 ships. In public works 16 fountains and pontoons and a sewer" (Salivia 1950, 194). Schwartz (2015, 38). Infrastructure: Deficient infrastructure contributed to losses: "84% of residences in Puerto Rico had the characteristics of poverty" (Caldera Ortíz 2014, 93). He speculates the storm contributed to the Lares insurrection in 1868 (2014, 93–94). Technology: "The first tropical cyclone where the wind speed is known" (Garcia).

1870s — Technology: In Cuba Father Benito Vines establishes observation sites and develops the first method for predicting hurricane movement.

1871, Aug. 21, hurricane Santa Juana — "Damage was minimal" (Garcia). Vivas (1962, 320). Technology and Infrastructure: Advance cable warning. "[In] 1869 Governor Sanz had built the first 95 km telegraph line between the capital and Arecibo and in 1870 there was telegraphic communication with Ponce. Submarine cable had been laid in 1870" (Salivia 1950, 203–204).

1872 (novel) — Alejandro Tapia y Rivera. *Póstumo el transmigrado* ("Posthumous the Transmigrated").

1873 — Technology: Father Benito Vines in Cuba issues the first know hurricane warning.

1876, Sept. 13, hurricane San Felipe I — Caldera Ortíz reports 19 dead, 91,221 pesos lost (2014, 99). Neely labels it Category 3 (2017, 228). "[O]ne of great cyclones of the nineteenth century, but it did less damage than Santa Ana, San Narciso and San Ciriaco. It was much inferior to San Felipe in 1928" (Salivia 1950, 223). Vivas (1962, 320). Technology: "The first storm where rain was measured in Puerto Rico. A total of 19 deaths was associated with the storm" (Garcia).

1876 (novel) — Alejandro Tapia y Rivera. *El pirata Cofresí* ("The Pirate Cofresí").

NOTE: Between 1876 and 1888 there is a gap in well-documented disasters.

1880 (memoirs) — Alejandro Tapia y Rivera. *Mis memorias* ("My Memoirs," posthumous).

1880 (poetry) — José Gautier Benítez. *Poesías* ("Poems").

1885 (poetry) — Lola Rodríguez de Tió. *Claros y nieblas* ("Clearings and Mists" 1885).

1888, Sept. 1–2, hurricane San Gil — Category 1, 100 deaths (Blake 2011). Example of a storm that is destructive despite not hitting the island directly, it passed 100–140 miles to the northwest. "This hurricane is similar to Donna in 1960 which had the same tragic effect in Puerto Rico without passing over the island" (Garcia). Miner Solá notes rains caused deaths (1995, 25). More than 100 people died from floods; "great losses of livestock, and agriculture suffered damage estimated at many thousands of pesos" (Salivia 1950, 228). Vivas places it on Sept. 15 labeling it San Lorenzo (1962, 320). Infrastructure: Rains caused destruction because of substandard construction and ineffectual planning.

1889, Sept. 4, hurricane San Moisés — Vivas (1962, 320).

1891, Aug. 19–20, hurricane San Magín — Extensive flooding (Boose). 8 deaths (Miner Solá 1995, 27). Vivas (1962, 320).

1893 Aug. 16, hurricane San Roque II — Coffee crop suffered major damages (Caldera Ortíz 2014, 106). 4 deaths (Miner Solá 1995, 28). Vivas (1962, 320). Technology: "[F]irst storm where flags were used to warn people of a storm coming, done in government offices" (Garcia).

1893 (painting) — Francisco Oller y Cestero. *El velorio* ("The Wake").

1894 (history) — Brau, Salvador. *Puerto Rico y su historia: investigaciones críticas.* ("Puerto Rico and Its History: Critical Research").

1894 (novel) — Manuel Zeno Gandía. *La charca* ("The Pond").

1897 — Technology: Telephone service is inaugurated.

1898–1899 (painting) — Francisco Oller y Cestero. *Hacienda Aurora.*

1899, Aug. 8, hurricane San Ciriaco, Earthquakes — Category 4. First hurricane after Puerto Rico became a territory of the United States. Crossed the entire length of the island (Neely 2017, 233). Caused 3369 deaths, "2,183 deaths more than those caused by all the other hurricanes that have occurred in Puerto Rico [until 1950]" (Salivia 1950, 255). Damage estimates: $35,889,013 (Blake 2011). Estimated sustained winds 140–145 mph (Garcia). A severe blow to the coffee industry (Mixer 66). "Towns such as Patillas, Maunabo and Yabucoa disappeared... Its buildings were destroyed almost entirely" Salivia states there were two strong earthquakes (Salivia 1950, 276). Caldera Ortíz comments on the general outlook before San Ciriaco: "82.4% illiteracy rate, few factories and considered primitive compared to those of Cuba or the United States, poor communication routes, and the economy in limbo due to currency devaluation and congressional debate over what status the island would have" (2014, 107). [T]he storm's symbolic weight: "Despite the promises of benevolence and munificence, many saw an ill omen for the occupation, when San Ciriaco devastated Puerto Rico on August 8, 1899, a little over a year after U.S. troops had landed" (Diebert 2019, 35). Middledyk says it devastated the intellectual scene, especially reading-rooms: "Arecibo, Vega-Baja, Toa-Alta, Yauco, Cabo-Rojo, Aguadilla, Humacao, and others made efforts in this direction either through their municipalities or private initiative. Only a few succeeded, but they did not outlive the critical times that commenced with the war, aggravated by the hurricane of August, 1898" (1903, 241). He must mean 1899 since no other source reports a hurricane in 1898, unless you count the *metaphorical* hurricane that was the U.S. invasion. Technology: Middledyk says there was a meteorological station in operation in San Juan: "rainfall was excessive, as much as 23 inches falling at Adjuntas during the course of twenty-four hours.... drowning deaths numbered 2,569 as compared with 800 killed by injuries received from the effects of the wind. This number does not include the thousands who have since died from starvation. The total loss of property was 35,889,013 pesos" (1903, 220).

He states that the U.S. provided aid: "the U.S. Government and people promptly came to the assistance of the starving population, and something like 32,000,000 rations were distributed by the army during the ten months after" (1903, 221). Vivas also mentions U.S. help: "U.S. aid was quick and effective as far as possible. Large quantities of clothing, food, and construction materials were brought to the island and distributed among the victims. This hurricane was the last one of the century and at the same time the definitive beginning of the trauma in which Puerto Rico is plunged" (1962, 184). He undermines his praise for aid with his mention of "trauma," presumably from the colonialism the island once again confronted.

NOTE: Between 1899 and 1916 there is a gap in well-documented disasters.

1904 (poetry) — José de Diego. *Pomarrosas* ("Rose Apple Trees").

1912 (painting) — Francisco Oller y Cestero. *Paisaje de Puerto Rico* ("Landscape of Puerto Rico").

1913 (poem) — Luis Lloréns Torres. "Canción de las Antillas" ("Song of the Antilles").

1914 (history) — Cayetano Coll y Toste. *Boletín histórico de Puerto Rico* ("Historical Bulletin of Puerto Rico" 1916).

1916, Aug. 22, [II] hurricane San Hipolito — Category 2, 1 death, $1,000 (damages), $36,000 (adjusted for inflation) (Blake 2011). Vivas (1962, 320).

1918, Earthquakes and Tsunami — "The earthquake changed natural environments, such as rivers" (Jaramillo Nieves 2018, 84). American Red Cross "provided great services in the earthquakes" (Salivia 1950, 316–317). 116 dead, $4 million in damages (Wagenheim 1970, 28).

1922 — Differential equations first used for forecasting hurricanes.

1926, July 23, hurricane San Liborio — Category 2, 25 deaths, $5,000 (damages) $103,353 (adjusted for inflation) (Blake 2011). Vivas (1962, 320).

1928, Sept. 13, hurricane San Felipe II [also known as Okeechobee for the Florida area it struck] — Category 5, 312 deaths, $85,000 (damages), $1,757,006 (adjusted for inflation). "The deadliest hurricane in Puerto Rico since 1899" (Blake 2011). Unprecedented damages: "some sugar processing *centrales* reduced to rubble" (Miner Solá 1995, 36). Many children under the age of ten died (Miner Solá 1995, 37). Neely says radio broadcasts gave early warning; several hundred thousand people homeless (2017, 272). Schwarz says a third of the population of 1.5 million were homeless; reports the American Red Cross helped recovery (2015, 41). Salivia says, "Despite being the most terrible of our hurricanes, both San Ciriaco and Santa Ana caused a greater number of deaths" (1950, 316); Vivas examines the colonial context: "During the late twenties, the United States suffered a great depression. Puerto Rico undoubtedly had to suffer the consequences. In addition, the island was punished by the hurricane of San Felipe in 1928. The winds of this hurricane reached speeds greater than 200 miles per hour, causing 312 deaths, thousands of injuries, a total loss of more than 85 million dollars and more than half a million people were left homeless. This hurricane destroyed the coffee plantations" (1962, 212). Technology: Radio warnings; Salivia compares the death total of 3000 in San Ciriaco to 300 for San Felipe II due to advance warning (1950, 321).

1929 (*plena* song) — "Temporal, temporal" ("Tempest, Tempest").

1931, Sept. 10, hurricane San Nicolás — Category 1, 2 deaths, $200 (damages) $4,386 (adjusted for inflation) (Blake 2011). Vivas (1962, 320). Infrastructure: Interruption of electrical service widely recorded for San Nicolás and progressively becomes a more devastating consequence of hurricanes in the twentieth century. Salivia gives more context:

> Three years after the formidable hurricane of San Felipe II, when the Jíbaro of the Fortaleza ruled the island, [Theodor Roosevelt, Jr. Governor from 1929-1932] and the people of Puerto Rico desperately struggled to recover from the tremendous damage caused by said hurricane, in addition to the economic crisis that arose in 1929 as a consequence of the First World War; there came the storm of September 10, 1931, known as San Nicolás, to aggravate the situation even more (1950, 329).

1932, Sept. 26, hurricane San Cipriano II or San Ciprián — Category 4, 225 deaths in PR and US Virgin Islands $30,000 (damages) $657,893 (adjusted for inflation) (Blake 2011). More than 25,000 families homeless (Miner Solá 1995, 38). Salivia narrates how he and his family survived this storm (1950, 336–338). "San Ciprián caused 257 deaths, 4280 injured, damages of more than 40 million dollars. The island was left defenseless by two catastrophes [first San Felipe in 1928] that added to the prevailing poverty and unemployment" (Vivas 1962, 212).

NOTE: Between 1932 and 1956 there is a gap in well-documented disasters.

1934 (essay) — Antonio S. Pedreira. *Insularismo* ("Insularism").

1935 (novel) — Enrique Laguerre. *La llamarada* ("The Blaze").

1940 (novel) — Enrique Laguerre. *Solar Montoya.*

1936 (stories) — Miguel Meléndez Muñoz. *Cuentos del cedro* ("Cedar Tales").

1937 (novel) — Manuel Méndez Ballester. *Isla cerrera* ("Untamed Island").

1937–1950 (poetry) — Luis Palés Matos. *Tuntún de pasa y grifería* ("Drumbeats of Kinkiness and Blackness").

1938 (poetry) — Julia de Burgos. *Poema en veinte surcos* ("Poem in Twenty Furrows").

1939 (essay) — Margot Arce de Vázquez. "El paisaje de Puerto Rico" ("The Landscape of Puerto Rico").

1939 (poetry) — Julia de Burgos. *Canción de la verdad sencilla* ("Song of the Simple Truth").

1940 (painting) — Miguel Pou Becerra. *Las calesas de Ponce* ("The Carriages of Ponce").

1944 — Technology: First hurricane hunter airplanes used which fly through the hurricane eye at more than 10,000 feet above the ocean.[7]

1947 (story) — Abelardo Díaz Alfaro. "El josco" ("The Strong One").

1947 (stories) — González, José Luis. *En la sombra* ("In the Shadow").

1935 (novel) — Enrique Laguerre. *La resaca* ("The Undertow").

1940 (tragedy) — Manuel Méndez Ballester. *Tiempo muerto. Tragedia en tres actos* ("Off Season. A Tragedy in Three Acts").

1950 (poetry) — Clara Lair. *Trópico amargo* ("Bitter Tropic").

1950 (story) — González, José Luis. "En el fondo del caño hay un negrito" ("At the Bottom of the Canal There Is a Little Black Boy").

1950s — Technology: Computer advances leveraged for statistical and dynamical modeling in hurricane forecasting.

1952 (poetry) — Francisco Matos Paoli. *Canto a Puerto Rico* ("Song for Puerto Rico").

1952–1954 (mural) — Rafael Tufiño. *La Plena* ("The Plena Song").

1939 (poetry) — Julia de Burgos. *El mar y tú y otros poemas* ("The Sea and You and Other Poems," posthumous).

1955 (engravings) — Lorenzo Homar and Rafael Tufiño. *Plenas: 12 grabados de Lorenzo Homar y Rafael Tufiño* ("*Plena* Songs: 12 Engravings by Lorenzo Homar and Rafael Tufiño").

1955 (stories) — Tomás Blanco. *Los cinco sentidos* ("The Five Senses").

1956 (play) — René Marqués. *La Carreta* (*The Oxcart*).

1956, Aug. 12, hurricane Santa Clara or Betsy — Category 2, 116 deaths, $40,000 (damages) $336,855 (adjusted for inflation) (Blake 2011). Technology: Radar for hurricane tracking (Miner Solá 1995, 40).

1960, Sept. 5, hurricane San Lorenzo or Donna — Category 4. PR & St. Thomas, 107 deaths (Blake 2011); 100 deaths (Miner Solá 1995, 1).

1961 — Technology: First hurricane-tracking satellite (Miner Solá 1995, 70).

NOTE: Between 1960 and 1975 there is a gap in well-documented disasters.

1962 (history) —José Luis Vivas. *Historia de Puerto Rico* ("History of Puerto Rico").

1960s — Technology: Unmanned aerial vehicles (UAVs) and drones for tracking.

1963 (poster) — José M. Figueroa. *Huracán* (Hurricane).

1964 — Technology: NOAA begins issuing three-day hurricane forecasts.

1965 (poster) — Eduardo Vera Cortés. *Huracán* ("Hurricane").

1965 (poem) — Juan Antonio Corretjer. "Alabanza en la torre de Ciales" ("Hymn of Praise in the Tower of Ciales").

1965 (print) — Lorenzo Homar. *Unicornio en la isla* ("Unicorn on the island").

1970 (song) — Rafael Cortijo y su Combo. "Noche de Temporal" ("Stormy Night").

1970s — Technology: hurricane hunter airplanes begin using on-board radars and deploying dropsondes within hurricane cores.

1974 (novella) — Edgardo Rodríguez Juliá. *La renuncia del héroe Baltasar* ("The Renunciation").

1975, Sept. 15, hurricane Eloise — 44 deaths (Blake 2011).

1976 (stories) — Rosario Ferré. *Papeles de Pandora* ("Pandora's Papers").

1976 (novel) — Luis Rafael Sánchez. *La guaracha del Macho Camacho* ("Macho Camacho's Beat").

1979, Aug. 3, hurricane David — Category 1, unknown deaths (Blake 2011). "Agricultural loss was estimated at $50 million. Coffee crop loss was 50% and 60% of the plantain crop was lost" (Boose).

1979, Sept. 4, hurricane Frederic — Category 4, 7 deaths $125,000 (damages) $357,143 (adjusted for inflation) (Blake 2011). Flooding throughout; $55 million coffee crop loss (Boose).

1980 (essay) — José Luis González. *El país de cuatro pisos* ("The Four-Storied Country").

1981 (stories) — Ana Lydia Vega and Carmen Lugo Filippi. *Vírgenes y mártires, cuentos* ("Virgins and Martyrs, Tales").

1982 (stories) — Ana Lydia Vega. *Encancaranublado y otros cuentos de naufragio* ("*Encancaranublado* and Other Tales of Shipwreck").

1984 (novel) — Edgardo Rodríguez Juliá. *La noche oscura del niño Avilés* ("The Dark Night of the Child Avilés").

1986 (history) — Fernando Picó, *Historia general de Puerto Rico* ("General History of Puerto Rico").

1987 (stories) — *Pasión de historia y otras historias de pasión* ("Passion for History and Other Stories of Passion").

1988 (novel) — Magali García Ramis, *Felices días, tío Sergio* ("Happy Days, Uncle Sergio").

1989, Sept. 18, hurricane Hugo — Category 3, 5 deaths (PR and US Virgin Islands), $1,000,000, $1,824,953 (adjusted for inflation) (Blake 2011). In Rio Piedras, "95% of the trees at the Botanical Garden had broken branches or were blown down. Many unique species, like "nuez de Macadamia," were seriously damaged... 2,000 orchids lost.... It will take thirty or forty years to regain its splendor"; "Radar out of service. Lots of airplanes destroyed.... The International Tower (telephone communication) blew down and the island was incommunicado. No air traffic for a few days" (Boose). "The major disaster related to this hurricane due to intense winds and excessive infrastructure [damage]" (Caldera Ortíz 2014, 136). "Many of our neighbors were spending time together in front of their houses. There was great expectation, because many had never seen a hurricane, especially the children... they didn't imagine they would be deprived of water and electricity for a long time"; Lake Carraízo dam threatened (Miner Solá 1995, 54).

1989 (stories) — Rosario Ferré. *Maldito Amor* ("Sweet Diamond Dust").

1991 (stories) — Ana Lydia Vega. *Falsas crónicas del sur* ("False Chronicles of the South").

1993 (essay) — Arcadio Díaz Quiñones. *La memoria rota.* ("The Broken Memory").

1995, Sept. 16, hurricane Marilyn — Category 3, 8 deaths (P.R. and U.S.V.I), $1,500,000 (damages) $2,254,601 (adjusted for inflation) (Blake 2011). Garcia notes that "Culebra had much severe damage as the eyewall blasted through... damage estimates comparatively close to those of hurricane Hugo in 1989."

1995 (novel) — Rosario Ferré. *The House on the Lagoon.*

1996, Sept. 10, hurricane Hortense — Category 1, 18 deaths $500,000 (damages) $737,952 (adjusted for inflation) (Blake 2011). "At 22 municipalities on the southeast side of the island, 650 houses suffered considerable damage. Agricultural loss was calculated at $127 million"; "About 40% of the island's coffee crop was blown down. A total of 60 telephone towers were blown down"; "Trees blown over by the winds made it difficult to reach some of the isolated communities on Vieques" (Boose).

1998, Sept. 21, hurricane Georges — Category 3, 0 deaths, $1,800,000 (damages) $2,512,821 (adjusted for inflation). Costliest hurricane to affect Puerto Rico until 2011 (Blake 2011); more than 30,000 homes were destroyed and 50,000 suffered minor damage; it also destroyed 75% of the coffee crop and 95% of plantation crops (Neely 2017, 330).

1999, Nov 17, hurricane Lenny — Category 3, 0 deaths, $330,000 (damages) $441,201 (adjusted for inflation) (Blake 2011).

2000, Aug. 22, hurricane Debbie — Category 1, 0 deaths, $25,000 (damage); passed about 30 to 40 miles off the northeast coast of Puerto Rico and caused heavy rains and much damage. NOAA says, "Some sectors in the interior received more than 12 inches of rain in a 48-hour period."[8]

2000 (essay) — Arcadio Díaz Quiñones. *El arte de bregar y otros ensayos* ("The Art of Coping and Other Essays").

2000 (novel) — Mayra Santos-Febres. *Sirena Selena vestida de pena* ("Sirena Selena").

2002 (essay) — Eduardo Lalo. *La isla silente* ("The Silent Island").

2004 (essay) — Eduardo Lalo. *La inutilidad* ("Futility").

2005 (essays) — Edgardo Rodríguez Juliá. *San Juan: Ciudad soñada* ("San Juan: Dreamt-of City").

2005 (novel) — Magali García Ramis. *Las horas del sur o la vida de Andrés Esterrich de quien se dijo que fue el primero en cantar la Canción Mixteca en las Antillas. Con testimonios de sus aventuras y vicisitudes y acotaciones de su desaparición* ("The Southern Hours or The Life of Andrés Esterrich Who Was Said to Be the First to Sing the Mixtec Song in The Antilles. With Testimonies of His Adventures and Vicissitudes and Notes of His Disappearance").

2005 (essay) — Eduardo Lalo. *Donde* ("Where").

2008 (chromogenic print) — Carlos Ruiz Valarino. *Sin título* ("Untitled").

2008 (essays) — Ana Lydia Vega. *Mirada de doble filo* (Double-Edged Gaze).

2010 (stories) — Luis Negrón. *Mundo Cruel* ("Cruel World").

2011, Aug. 22, hurricane Irene — Category 1, 0 deaths, $5.000 million (damages). Made landfall in Puerto Rico as a tropical storm, before strengthening to a Category 4 over the main island.

2011 (stories) — Janette Becerra. *Doce versiones de soledad.*

2016 (graphic novel) — Edgardo Miranda-Rodríguez. *La Borinqueña.*

2017, Sept. 20, hurricane Maria — Category 4. Estimated 2,975[9]–4,000[10] deaths; $90,000,000,000 (damages to U.S.V.I. and P.R.).[11] Deadliest hurricane since 1899's San Ciriaco. NOAA categorizes the damage as catastrophic.[12] Diebert offers revealing details: "In the south eastern town of Yabucoa, near where the storm made landfall with winds reaching 165 mph, 99% of the municipal buildings collapsed" (2019, 114). NOAA says San Felipe II had been the last Category 4 or stronger hurricane.[13] Maria hit when the archipelago's electrical infrastructure suffered from long-standing mismanagement and neglect. Provoked accelerated migration to the United States so that almost a quarter million people left. As a result, everything from everyday services to medical care worsened. The Trump administration, FEMA and local bureaucracies impeded use of billions of Congress-apportioned dollars so that when the next big storm hit — Fiona in 2022, even though it was a lowly Category 1, the island was once again plunged into darkness.

2017–2022 (5 paintings) — Enoc Pérez. *Paisajes* ("Landscapes").

2018 (history) — Lorna Gisela Jaramillo Nieves, *11 de octubre de 1918. El terremoto en Puerto Rico. Lecciones cien años después.* ("October 11, 1918. The Earthquake in Puerto Rico. Lessons a Hundred Years Later").

2018 (graphic novel) — Edgardo Miranda-Rodríguez (producer), *Ricanstruction.*

2018 (poetry) — Mayra Santos-Febres, *Huracanada* ("Hurricaned").

2018–2020 (Assemblage) — Daniel Lind-Ramos. *Maria Guabancex.*

2019 (silkscreen series) — Guillermo E. Rodríguez. *Doppler Landscapes.*

2019 (documentary) — Nadia Hallgren. *After Maria.*

2020, Jan. 29, 6.4 magnitude earthquake and aftershocks — "The Puerto Rico Seismic Network registered 43 'significant' quakes, leaving many people fearful of going home. More than 8,000 houses damaged."[14] "About 7,500 people left their homes for other shelter"; "aftershocks will persist for years to decades, although with decreasing frequency."[15]

2020, March (beginning) — Covid-19 pandemic. The Johns Hopkins University of Medicine Covid-19 data center kept track for three years and there were 1,101,469 confirmed cases and 5,823 deaths.[16] Because vaccines and masking were not politicized, Puerto Rico mitigated the epidemic successfully.[17]

2020 (Assemblage) — Daniel Lind-Ramos. *Ambulancia* ("Ambulance").

2020 (memoir) — Mayra Santos-Febres, *Antes que llegue la luz* ("Before the Electricity Returns").

2020 (Assemblage) — Daniel Lind-Ramos. *Baño de María. (Bain-Marie/The Cleansing).*

2020 (painting) — Jotham Malavé. *El bautismo colectivo* ("Collective Baptism").

2022, Sept. 18, hurricane Fiona — Category 1. Infrastructure: Puerto Rico's infrastructure suffered 25–30 inches of rain in some areas. The entire island lost power and water. *The New York Times* reported on Nov 11, 2022, that 25 people died.[18]

2022 (song) — Bad Bunny. "El apagón" ("The Blackout").

2022 (documentary) — Bianca Graulau. "Aquí vive gente" ("People Live Here").

2023 (paitings) — Enoc Pérez. *Paisajes.* ("Landscapes").

2022–2023 — Whitney Museum exhibit *No existe un mundo poshuracán: Puerto Rican Art in the Wake of Hurricane Maria.*

2023 (play) — Isabel Ramos and Kisha Tikina Burgos. *Antígona frente al mar* ("Antigone Before the Sea").

Notes

1. https://web.archive.org/web/20250210141122/https://www.weather.gov/mob/tropical_events
2. "Decline in North Atlantic major hurricane frequency in the twentieth century was particularly notable in the 1970s and 1980s and unprecedented in one 270-year reconstruction" (Chenoweth and Howard 2023, 11).
3. Bletery and Nocquet analyzed high-rate GPS time series before magnitude 7 > earthquakes looking for precursor signals and observed a subtle signal from the noise about 2 hours before the earthquakes that in future may allow monitoring of faults during this precursor phase with denser and higher-precision instrumentation.
4. Editor José F. Buscaglia-Salgado proves Ramírez was historical in the Rutgers UP 2018 edition.
5. *The Great Hurricane of 1780* https://www.nhc.noaa.gov/pastdeadlyapp2.shtml?

6. https://issuu.com/shpopr/docs/11_issuu_la_ciudad_en_el_tiempo
7. http://web.archive.org/web/20250210142728/https://www.washingtonpost.com/news/capital-weather-gang/wp/2014/09/29/70-years-ago-hurricane-hunters-got-their-start-in-the-great-atlantic-hurricane-of-1944/
8. https://web.archive.org/web/20250210143321/https://www.ncdc.noaa.gov/stormevents/eventdetails.jsp?id=5156116
9. The George Washington University. "Ascertainment of the Estimated Excess Mortality from Hurricane María in Puerto Rico." https://web.archive.org/web/20250221113822/https://publichealth.gwu.edu/sites/g/files/zaxdzs4586/files/2023-06/acertainment-of-the-estimated-excess-mortality-from-hurricane-maria-in-puerto-rico.pdf
10. *The New York Times*. "Puerto Rico's Hurricane Maria Death Toll Could Exceed 4,000, New Study Estimates". https://web.archive.org/web/20250210144541/https://www.nytimes.com/2018/05/29/us/puerto-rico-deaths-hurricane.html
11. https://web.archive.org/web/20250209180144/https://www.nhc.noaa.gov/news/UpdatedCostliest.pdf
12. https://web.archive.org/web/20250210143758/https://www.ncdc.noaa.gov/stormevents/eventdetails.jsp?id=721909
13. https://web.archive.org/web/20250210143758/https://www.ncdc.noaa.gov/stormevents/eventdetails.jsp?id=721909.This is the database devoted to Maria exclusively. Contains all the information NOAA has on the storm.
14. https://web.archive.org/web/20250210151550/https://www.nytimes.com/2020/03/01/us/puerto-rico-earthquakes-fema.html
15. https://web.archive.org/web/20250210152303/https://www.usgs.gov/news/featured-story/magnitude-64-earthquake-puerto-rico
16. https://web.archive.org/web/20240920074906/https://coronavirus.jhu.edu/region/us/puerto-rico
17. https://web.archive.org/web/20250210153653/https://www.aha.org/news/blog/2022-07-22-digging-reasons-puerto-ricos-successful-covid-19-response
18. https://web.archive.org/web/20250210154451/https://www.nytimes.com/2022/11/08/magazine/puerto-rico-farms-hurricanes.html

Bibliography

Abbad y Lasierra, Fray Iñigo. *Historia geográfica, civil y natural de la isla de la isla de San Juan Bautista de Puerto Rico*. 1788. José Julián de Acosta y Calbo, editor. Imprenta y Librería de Acosta, 1866. https://web.archive.org/web/20250310131442/https://www.cervantesvirtual.com/obra/historia-geografica-civil-y-natural-de-la-isla-de-san-juan-bautista-de-puerto-rico--por-inigo-abbad-y-lasierra/.

Abram, Christopher. *Evergreen Ash. Ecology and Catastrophe in Old Norse Myth and Literature*. Virginia UP, 2019.

Acosta Cruz, María. *Dream Nation: Puerto Rican Culture and the Fictions of Independence*. Rutgers UP, 2014.

Acosta y Calbo, José Julián de. Section on hurricanes. Abbad y Lasierra, *Historia geográfica*: 433–435. https://archive.org/details/historiageografi00abba/page/n6/mode/1up?ref=ol&view=theater.

Alaimo, Stacy. "Trans-corporeal Feminisms, and the Ethical Space of Nature." *Material Feminisms*. Stacy Aliamo and Susan Hekman, editors, 2008: 237–264.

Alaimo, Stacy and Susan Hekman, editors. *Material Feminisms*. Indiana UP, 2008.

Aldama, Frederick Luis. *Latinx Superheroes in Mainstream Comics*. Arizona UP, 2017.

Alonso, Manuel A. (Manuel Antonio). *El gíbaro*. Barcelona, 1849. https://www.archive.org/details/elgibaro00alon/mode/1up

Alonso Manuel A., Santiago Vidarte et al. *Cancionero de Borinquen*. Imprenta de Martí Carlé, 1846. https://www.google.com/books/edition/El_Cancionero_de_Borinquen/0ZkQAAAAIAAJ?hl=en.

M. Acosta Cruz, *An Ecocritical Study of Puerto Rican Culture*, https://doi.org/10.1007/978-3-032-04044-2

Aponte Alsina, Marta. "Los códigos de la isla." *Plural*, vol. 3, no. 1–2 (enero-diciembre 1984): 89–95.

Aponte Alsina, Marta, editor. *Narraciones puertorriqueñas*. Biblioteca Ayacucho, 2015.

Anderson, Mark D. *Disaster Writing. The Cultural Politics of Catastrophe in Latin America*. Virginia UP, 2011.

Anonymous, "Letter of complaint against the colonial regimes of don Gaspar de Arteaga (1670–1674) and don Juan de Robles Lorenzana (1678–1683)." Rivera de Álvarez, *Literatura*, p. 37, note 12.

Arons, Wendy. "Tragedies of the Capitalocene." *JCDE*, vol. 8, no. 1: 16–33.

Arroyo, Jossiana. *Caribes 2.0. New Media, Globalization, and the Afterlives of Disaster*. Rutgers UP, 2023.

Arroyo Pizarro, Yolanda. "De huracanes y corazones torcidos..." Blog post, 2006. https://web.archive.org/web/20250310131531/https://narrativadeyolanda.blogspot.com/2006/08/de-huracanes-y-corazones-torcidos.html

Asenjo, Federico. *Efemérides de la Isla de Puerto Rico*. Imp. de J. Gonzalez Font San Juan, P.R 1886. https://harvardforest1.fas.harvard.edu/sites/harvardforest.fas.harvard.edu/files/data/p01/hf012/hf012-05-reports.txt.

Arce de Vázquez, Margot. "El paisaje de Puerto Rico." *Literatura puertorriqueña del siglo XX: Antología*. Mercedes López-Baralt, editor: pp. 68–75.

Avilés-Santiago, Manuel G. "The Technological Embodiment of Colonialism in Puerto Rico." *Anthurium: A Caribbean Studies Journal*, vol. 12, iss. 2, article 6, 2015. https://web.archive.org/web/20250302211430/ https://anthurium.miami.edu/articles/10.33596/anth.294.

Babín, María Teresa. *Fantasía Boricua. Estampas de mi tierra*. Instituto de Cultura Puertorriqueña, 1973.

Bagué, Katiana. "Revisitando huracanes del pasado en Puerto Rico." *The US Caribbean & Ethnic Florida Digital Newspaper Project*, 2020. https://web.archive.org/web/20230123181846/https://ufndnp.domains.uflib.ufl.edu/revisitando-huracanes-del-pasado-en-puerto-rico/.

Becerra, Janette. *Doce versiones de soledad*. Ediciones Callejón, 2011.

Bentley, Chris. "Puerto Rico Hopes Solar Project Will Secure Electric Grid for Future Hurricanes." 17 Sept. 2023. https://web.archive.org/web/20250302210850/https://www.npr.org/2023/09/17/1200053507/puerto-rico-hopes-solar-project-will-secure-electric-grid-for-future-hurricanes.

Bergthaller, Hannes. "Limits of Agency. Notes on the Material Turn from a Systems–Theoretical Perspective." Serenella Iovino and Serpil Oppermann, editors. *Material Ecocriticism*. Indiana UP (2014): 37–50.

Bermúdez Pérez, Adrián. "From Ruined to Rebellion: Hurricanes in El Grito de Lares and El verano del 2019 in Puerto Rico." Instituto de Estudios del Caribe. YouTube, 3 Nov. 2022.

Blackmore, Lisa and Liliana Gómez. "Beyond the Blue Notes on the Liquid Turn." Lisa Blackmore and Liliana Gómez, editors. *Liquid Ecologies in Latin American and Caribbean Art*. Routledge, 2020.

Blake, Eric S. and Christopher W. Landsea. "The Deadliest, Costliest, and Most Intense United States Tropical Cyclones from 1851 to 2010 (And Other Frequently Requested Hurricane Facts)." *National Oceanic and Atmospheric Administration* (NOAA) Technical Memorandum NWS NHC-6. National Weather Service National Hurricane Center, August 2011, https://web.archive.org/web/20250310133433/https://repository.library.noaa.gov/view/noaa/6929.

Blanco, Tomás. *Los cinco sentidos*. Instituto de Cultura Puertorriqueña, 1968.

Bletery, Quentin and Jean-Mathieu Nocquet. "The Precursory Phase of Large Earthquakes." *Science*, vol. 381, no. 6655: 297–301. https://web.archive.org/web/20250310133430/ https://www.science.org/doi/10.1126/science.adg2565.

Bonilla, Yarimar and Marisol LeBrón, editors. *Aftershocks of Disaster: Puerto Rico Before and After the Storm*. Haymarket Books, 2019.

Boose, Emery. "Landscape and Regional Impacts of Hurricanes in Puerto Rico 1508–1997." https://web.archive.org/web/20250310133427/https://harvardforest1.fas.harvard.edu/exist/apps/datasets/showData.html?id=HF012.

Borges-Méndez, Ramón and Cynthia Caron. "Decolonizing Resilience: The Case of Reconstructing the Coffee Region of Puerto Rico after Hurricanes Irma and Maria." *Journal of Extreme Events*, vol. 06, no. 01 (2019): 1940001-1-1940001-19.

Brau, Salvador. *Historia de Puerto Rico*. Appleton y Compañía, 1917. https://www.google.com/books/edition/Historia_de_Puerto_Rico/UXpsAAAAMAAJ?hl=en.

Brau, Salvador. *Lo que dice la historia*. 1893. www.gutenberg.org/files/42321/42321-h/42321-h.htm.

Brau, Salvador. *Puerto Rico y su historia: investigaciones críticas*. Imprenta de Francisco Vives Mora, 1894. https://archive.org/details/puertoricoysuhis00brau/page/n4/mode/1up?q=Puerto+Rico+historia.

Braun, Erick et al. "A 3000-yr Paleostorm Record from St. Catherines Island, Georgia." *Estuarine, Coastal and Shelf Science*, vol. 196 (September 2017): 360–372. https://web.archive.org/web/20250310134340/https://www.sciencedirect.com/science/article/pii/S0272771416303730?via%3Dihub&__cf_chl_rt_tk=SSJOXjn.n2JAwG1c_yVvPxD.tCE5ujZqYiNUGbqS7Vo-1741614220-1.0.1.1-ORoQqp0LSNmrBP5mjKQ_kSXpQruVYIWlHzKwxPRj.dk.

Brokaw, Galen. "Homo Alphabeticus, Glottographic Exceptionalism, and the Ethnocentric Definition of Writing." *PMLA*, vol. 137, no. 1, (2022): 163–170. https://doi.org/10.1632/S0030812922000013

Buell, Lawrence. *The Environmental Imagination. Thoreau, Nature Writing, and the Formation of American Culture*. The Belknap Press of Harvard UP, 1995.

Bunny, Bad. "El apagón." posted May 6, 2022, by Bad Bunny channel, YouTube, 3.59, https://www.youtube.com/watch?v=QdQEljUMCEM

Bunny, Bad. "Una velita," posted Sep 19, 2024, by Bad Bunny channel, YouTube, 3.59, https://www.youtube.com/watch?v=On1mm8vWJ50

Burgos, Julia de. *Song of the Simple Truth. Obra completa poética. The Complete Poems.* Compiled and translated by Jack Agüeros. Curbstone Press, 1997.

Byerly, Alison. "The Uses of Landscape. The Picturesque Aesthetic and the National Park System." *The Ecocriticism Reader: Landmarks in Literary Ecology*, Cheryll Glotfelty and Harold Fromm, editors. U of Georgia P (1997): 52–68.

Byron, Lord George Gordon. *Childe Harold's Pilgrimage*. Cassell & Company Limited, 1894. https://www.google.com/books/edition/Childe_Harold_s_Pilgrimage/dy8-AAAAYAAJ?hl=en&gbpv=0.

Caldera Ortiz, Luis. *La historia de los ciclones y huracanes tropicales en Puerto Rico*. Editorial Akelarre, 2014.

Caldwell, Nicole. "10 Ways Nature and Animals Forecast the Weather." 12 Aug. 2019, https://web.archive.org/web/20250310140009/https://stacker.com/stories/weather/10-ways-nature-and-animals-forecast-weather.

Caldwell, Wilber W. *American Narcissism: The Myth of National Superiority.* Algora Publishing, 2006.

Calley Galitz, Kathryn. "Romanticism." *Heilbrunn Timeline of Art History*. The Metropolitan Museum. https://web.archive.org/web/20250310140014/https://www.metmuseum.org/essays/romanticism.

Caño Martín Peña CLT/Fideicomiso de la Tierra. 29 Feb. 2016. Video. https://web.archive.org/web/20250310140106/https://www.youtube.com/watch?v=mNbjXzgnR88.

"Capetillo, Luisa." Valle Ferrer, Norma. Luisa Capetillo: obra completa. Puerto Rico: Departamento del Trabajo y Recursos Humanos, 2008.

Caro González, Leysa. "No arrancan obras en las escuelas afectadas por los terremotos." *El Nuevo Día*, 6 Jan. 2021. https://web.archive.org/web/20250310140650/https://www.elnuevodia.com/noticias/locales/notas/no-arrancan-obras-en-las-escuelas-afectadas-por-los-terremotos/.

Caruth, Cathy. *Unclaimed Experience. Trauma, Narrative, and History*. The Johns Hopkins UP, 1996.

Castellanos, Juan de. *Elegías de varones ilustres de Indias*. Biblioteca de Autores Españoles, 1847. https://web.archive.org/web/20250310140656/https://bibliotecadigital.jcyl.es/i18n/catalogo_imagenes/grupo.do?path=10076843.

Chakrabarty, Dipesh. "The Climate of History: Four Theses." *Critical Inquiry*, vol. 35, no. 2 (Winter 2009): 197–222.

Cheng, Lijing et al. "Past and Future Ocean Warming." *Nature Reviews*, vol. 3 (Nov. 2022): 776–794.

Chenoweth, Michael and Ian Howard. "Hurricane Impacts on Land in the Central and Eastern Caribbean since 1494 CE from Written Records." *Earth and Space Science*, vol. 10. iss. 7, July 2023. https://web.archive.org/web/20250310140645/, https://agupubs.onlinelibrary.wiley.com/, https://doi.org/10.1029/2023EA002897.

Clark, Timothy. "Derangements of Scale." *Telemorphosis: Essays in Critical Climate Change*, Tom Cohen and Henry Sussman, editors. Vol. i (2012): 148–66.

Coleridge, Samuel Taylor. *The Rime of the Ancient Mariner*. Globe School Book Company, 1900. https://www.google.com/books/edition/The_Rime_of_the_Ancient_Mariner/pHURAAAAYAAJ?hl=en&gbpv=0

Coll y Toste, Cayetano. *Boletín histórico de Puerto Rico.* Puerto Rico, Ateneo Puertorriqueño, Editorial LEA, 1916. https://web.archive.org/web/20250310141649/https://catalog.hathitrust.org/Record/006791914.

Colón, Cristóbal. "Carta a Luis de Santángel." 1493. https://web.archive.org/web/20250310141534/https://aix1.uottawa.ca/~jmruano/colon.pdf.

Colón Morales, Rubén. "Un país de escombros." 2 Oct. 2017, 80grados.net/un/pais-de-escombros/. https://web.archive.org/web/20250310141537/https://www.80grados.net/web/20250310141537/https://www.80grados.net/un-pais-de-escombros/.

Concepción, C.M. "El conflicto ambiental y su potencial hacia un desarrollo alternativo: el caso de Puerto Rico." *Ambiente y Desarrollo* 4.1, 1988: 125–35.

Condé, E.G. *Sordidez*. Stelliform Press, 2023. Kindle edition.

Cooper, Jago. "Fail to Prepare, Then Prepare to Fail: Rethinking Threat, Vulnerability, and Mitigation in the Precolumbian Caribbean." *Surviving Sudden Environmental Change. Answers from Archaeology*, edited by Jago Cooper and Payson Sheets. UP of Colorado, 2012. https://web.archive.org/web/20250310142548/https://www.jstor.org/stable/j.ctt1wn0rbs.9?seq=1.

Crutzen, Paul J. "Geology of mankind." *Nature*, vol. 415, iss. 6867 (2002). https://web.archive.org/web/20250310142703/https://www.nature.com/articles/415023a

Daniels, Stephen and Hayden Lorimer. "Until the End of Days: Narrating Landscape and Environment." *Cultural Geographies*, vol. 19, no. 1: 3–9. https://www.jstor.org/stable/44251449

Dávila, Tere. "Génesis." 2021. https://web.archive.org/web/20250325162918/https://historiasdeunacamara.com/2021/06/06/genesis/

Davis, Mike. *Ecology of Fear. Los Angeles and the Imagination of Disaster*. Vintage Books, 1999.

Delano, Paul. *The Museum of the Old Colony.* https://web.archive.org/web/20250310143509/https://artsandculture.google.com/asset/museum-of-the-old-colony-pablo-delano/aAGzMhUvdLhrRA?hl=en

DeLoughrey, Elizabeth. "Submarine Futures of the Anthropocene." *Comparative Literature*, vol. 69, no. 1 (2017): 32–44.

DeLoughrey, Elizabeth and Tatiana Flores. "The Tidalectics of Representability and the Sea in Caribbean Art". Lisa Blackmore and Liliana Gómez, editors. *Liquid Ecologies in Latin American and Caribbean Art*. Routledge (2020): 162–186.

Deopersad, Chitralekha and Luis Gabriel Azevedo. "Nubes de polvo del Sahara en el Caribe: peligro para los humanos y los arrecifes de coral." https://web.archive.org/web/20250322134106/https://idbinvest.org/es/blog/cambio-climatico/nubes-de-polvo-del-sahara-en-el-caribe-peligro-para-los-humanos-y-los

Díaz Quiñones, Arcadio. *El arte de bregar. Ensayos*. Ediciones callejón, 2000.

Díaz Quiñones, Arcadio. *La memoria rota*. Ediciones Huracán, 1993.

Diebert, Michael. *When the Sky Fell. Hurricane Maria in the United States in Puerto Rico*. Apollo Publishers, 2019.

Diego, José de. *Pomarrosas*. Imprenta de Heinrich y Co., 1904. https://www.google.com/books/edition/Pomarrosas/By1EAQAAMAAJ?hl=en

Dzieza, Josh. "When Will Puerto Rico Have Power?" 26 Oct. 2022. https://web.archive.org/web/20250218035644/https://www.theverge.com/c/features/23404252/puerto-rico-hurricane-maria-electric-grid-solar-panels-prepa-luma-fema.

Duany, Jorge. "Más allá de la docilidad. La antropología psicológica en Puerto Rico." Universidad del Sagrado Corazón. 1991.

Duany, Jorge. *Puerto Rico. What Everyone Needs to Know*. Oxford UP, 2017.

Ellis, Erle et al. "Involve Social Scientists in Defining the Anthropocene." *Nature*, vol. 540, no. 07 (Dec. 2016): 192–193. https://web.archive.org/web/20250310144151/https://www.nature.com/articles/540192a.

Emanuel, Kerry. *Divine Wind: The History and Science of Hurricanes*. Oxford UP, 2005.

Estok, Simon C. "Theorizing in a Space of Ambivalent Openness: Ecocriticism and Ecophobia." *Interdisciplinary Studies in Literature and Environment*, vol. 16, no. 2 (spring 2009): 203–225. https://web.archive.org/web/20250310144157/https://www.jstor.org/stable/44733418?seq=1.

Estok, Simon C. "Narrativizing Science: The Ecocritical Imagination and Ecophobia." *Configurations*, vol. 18, 2010, pp.141–159.

Estok, Simon C. "Painful Material Realities, Tragedy, Ecophobia." Serenella Iovino and Serpil Oppermann, editors. *Material Ecocriticism*. Indiana UP, 2014): 130–140.

Fan, Daidu and Liu, Kam-biu. "Perspectives on the Linkage between Typhoon Activity and Global Warming from Recent Research Advances in Paleotempestology." *Chinese Science Bulletin,* vol. 53 (2008): 2907–2922. https://web.archive.org/web/20250310144344/https://link.springer.com/article/10.1007/s11434-008-0341-2.

Fernández de Oviedo y Valdés, Gonzalo. *Historia general y natural de las Indias, islas y tierra-firme del mar océano.* Primera parte, 1535. https://web.archive.org/web/20250310145217/https://ems.kcl.ac.uk/content/etext/e026.html.

Ferré, Rosario. *Papeles de Pandora.* Vintage, 2000.

Ferré, Rosario. *The House on the Lagoon: A Novel.* Open Road Media, 2014.

Figueroa, Víctor. "Manos sucias, árboles y cemento: una lectura ecocrítica de *Huracanada* de Mayra Santos-Febres." *Latin American Literary Review*, vol. 48, no. 95 (2021): 9–16.

Flores, William. *Ecocrítica poscolonial y literatura moderna latinoamericana.* Universidad Nacional Mayor de San Marcos Fondo Editorial, 2015.

Foucault, Michel. *The Order of Things.* Random House, 1970.

Fuster Lavín, Ana María. "Arrojados al silencio. Microcuento." 2 Nov. 2017. https://web.archive.org/web/20250310145242/https://www.youtube.com/watch?v=KvSOD5JqESk

Gallardo, Frances. *Hurricane Series.* https://web.archive.org/web/20250310145732/https://www.francesgallardo.com/hurricanes.

Garcia, Jose M. "Hurricanes and Tropical Storms in Puerto Rico from 1980 to 2005." https://web.archive.org/web/20250322124728/https://huracanado1.tripod.com/history3.html

García-Calderón, Myrna, "La añoranza histórica en las obras de Magali García Ramis." *Explicación de Textos Literarios.* vol. 23, iss. 2 (1994): 55–63.

García-Moreno Barco, Francisco. "Patrimonio histórico cultural y desarrollo sostenible." *Verdor que clama,* Carmen M. Rivera Villegas et al, editors. Ediciones del Flamboyán, 2022: 145–152.

García Ramis, Magali. *Felices días, tío Sergio.* Editorial Santillana, 1986.

García Ramis, Magali. "Women's Tales." Asela Rodríguez de Laguna, editor. *Images and Identities: The Puerto Rican in Two World Contexts.* Routledge (1987): 109–115.

García Ramis, Magali. *La familia de todos nosotros.* Editorial Cultural, 1976.

García Ramis, Magali. *La R de mi padre y otras letras familiares.* Libros El Navegante, 2011.

García Ramis, Magali. *Las horas del sur.* Ediciones Callejón, 2005.

Garrard, Greg. *Ecocriticism.* 2nd ed. Routledge, 2012.

Gautier Benítez, José. "Puerto Rico." Rivera de Álvarez, Josefina and Manuel Álvarez Nazario, editors. *Antología general de la literatura puertorriqueña. Volume I.* Ediciones Partenón, (1982): 167–169.

Gerhardt, Christina. *Sea Change. An Atlas of Islands in a Rising Sea*. U of California P, 2023.

Guerrero, Marcela. *No existe un mundo poshuracán: Puerto Rican Art in the Wake of Hurricane Maria*. Whitney Museum of Art, 2022.

Glotfelty, Cheryll and Harold Fromm, editors. *The Ecocriticism Reader: Landmarks in Literary Ecology*. Georgia UP, 1997.

Gómez Rivera, Manuel. *José Pérez Losada: Vida y obra*. Repositorio UPR, 1969. https://web.archive.org/web/20250310145905/https://repositorio.upr.edu/.

Góngora y Sigüenza, Carlos de. *Infortunios que Alonso Ramírez, natural de la ciudad de San Juan de Puerto Rico, padeció, así en poder de ingleses piratas que lo apresaron en las islas Filipinas, como navegando por sí solo y sin derrotero, hasta parar en las costas de Yucatán: consiguiendo por este medio dar vuelta al mundo*. archive.org/details/infortuniosdealo00siguoft/page/28/mode/1up?ref=ol&view=theater.

Gonzalez, David. "A New Museum for an Old Colony, Puerto Rico." *The New York Times*, 15 Nov. 2017, https://web.archive.org/web/20250220200019/https://archive.nytimes.com/lens.blogs.nytimes.com/2017/11/15/a-new-museum-for-an-old-colony-puerto-rico/.

González, José Luis. "En el fondo del caño hay un negrito." https://web.archive.org/web/20221122205332/https://ciudadseva.com/texto/en-el-fondo-del-cano-hay-un-negrito/.

González, José Luis. *Cuentos completos*. Alfaguara, 1997.

González, José Luis. *Literatura y Sociedad en Puerto Rico*. Fondo de Cultura Económica, 1976.

Graulau, Bianca. *Aquí vive gente*. Documentary, 2022. YouTube.

Grove, Richard. *Green Imperialism: Colonial Expansion, Tropical Island Edens and the Origins of Environmentalism, 1600–1860*. Cambridge UP, 1995.

Hallgren, Nadia. *After Maria*. Film, 2019. Netflix.

Hartman, Joseph R. "Temporal Visions: Hurricanes as Chronotopes in Caribbean Art and Cultural Politics." Special issue *Decolonial Theory, Transculturation, and Latin American Positions – Entangling Art Histories*, Miriam Oesterreich and Franziska Koch, editors. *MIRADAS—Journal for the Arts and Culture of the Américas and the Iberian Peninsula* 7 (2023): 122–147, doi.org/https://doi.org/10.11588/mira.2023.1.96492

Herrera y Tordesillas, Antonio de. *Historia general de los hechos de los Castellanos en las islas, i tierra-firme del Mar Océano. 1492–1531*. Published 1601–1615. https://web.archive.org/web/20250310150427/https://www.cervantesvirtual.com/obra/historia-general-de-los-hechos-de-los-castellanos-en-las-islas-i-tierra-firme-del-mar-oceano-3/.

Hostos, E.M de. *La peregrinación de Bayoán*. 1863. Obras Completas, volumen VIII. Edición conmemorativa del gobierno de Puerto Rico, Cultural,

1839, 1939. https://web.archive.org/save/https://issuu.com/coleccionpuertorriquena/docs/la_peregrinaci_n_de_bayo_n_diario_regodido_y_publ.

Hulme, Mike. "Geographical Work at the Boundaries of Climate Change." *Transactions of the Institute of British Geographers* vol. 33, no. 1 (2008): 5–11.

Humphrey, Paul. "Framing a Decolonial Future: Hurricane María in Independent Puerto Rican Comics." *Latin American Literary Review*, vol. 48, no. 96 (summer 2021): 61–74.

Hung, Ruyu. "Towards Ecopedagogy: An Education Embracing Ecophilia." *Educational Studies in Japan: International Yearbook*, No. 11 (March 2017): 43–56.

Hunt, John Dixon. *Gardens and the Picturesque: Studies in the History of Landscape Architecture.* MIT Press, 1992.

Ilner, Peer and Isak Winsel Holm. "The Cultural Studies of Disaster." *Disaster Research. Multidisciplinary and International Perspectives*, Rasmus Dahlberg et al, editors. Routledge, 2016.

Instituto de Cultura Puertorriqueña. Expedientes COP. Colección de Cultura y Escultura. Database. Consulted March, 2023.

Instituto de Cultura Puertorriqueña. *Pintura puertorriqueña.* Colección de Pinturas del Instituto de Cultura Puertorriqueña. Programa de Artes Plásticas. Unidad de Colecciones. No date. ISBN 978-0-86581-621-3

Iovino, Serenella and Serpil Oppermann, editors. *Material Ecocriticism.* Indiana UP, 2014.

Janson, H.W. *History of Art: A Survey of the Major Visual Arts from the Dawn of History to the Present Day.* Prentice-Hall, 1977.

Jaramillo Nieves, Lorna Gisela. *11 de octubre de 1918. El terremoto en Puerto Rico. Lecciones cien años después.* Publicaciones Puertorriqueñas, 2018.

Kierner, Cynthia A. *Inventing Disaster: The Culture of Calamity from the Jamestown Colony to the Johnstown Flood.* U of North Carolina P, 2019. https://web.archive.org/web/20250310151306/https://www.jstor.org/stable/10.5149/9781469652535_kierner.3?seq=1.

Kishore, Natasha, at al. "Mortality in Puerto Rico after Hurricane Maria," *New England Journal of Medicine,* no. 379, 12 July 2018, pp.162–170. https://doi.org/10.1056/NEJMsa1803972

Klein, Naomi. *The Shock Doctrine: The Rise of Disaster Capitalism.* Picador, 2008.

Korten, Tristram. "The Bahamas and the Caribbean Have Withstood Hurricanes for Centuries." *Smithsonian Magazine*, 17 September 2019. https://web.archive.org/web/20250310151255/https://www.smithsonianmag.com/history/bahamas-and-caribbean-have-withstood-hurricanes-centuries-180973157/?__cf_chl_rt_tk=CptffThMd2L9kaxcMhgvlWwVd2mdQiNi39E9RdtPY6w-1741619575-1.0.1.1-91toYkc800WABVPb5AHqxP7OO7iYzrHkJa3k4GWV1DE

Kueffer, Christof and Kealohanuiopuna Kinney. "What is the Importance of Islands to Environmental Conservation?" *Environmental Conservation* 44, no. 4 (2017): 311–322 C.

La Fountain-Stokes, Lawrence. "Tomboy Tantrums and Queer Infatuations: Reading Lesbianism in Magali García Ramis's *Felices días, tío Sergio.*" *Tortilleras: Hispanic and U. S. Latina Lesbian Expression*, Lourdes Torres and Inmaculada Pertusa, editors, Temple UP; (2003): 47–67.

La Torre Lagares, Elidio. *Wonderful Wasteland and Other Natural Disasters: Poems*. Kentucky UP, 2019.

Las Casas, Fray Bartolomé de. *Brevísima relación de la destrucción de las Indias*. https://web.archive.org/web/20250220202522/https://aix1.uottawa.ca/~jmruano/relacion.pdf.

Las Casas, Fray Bartolomé de. *Historia de las Indias*. Edited by José Miguel Martínez Torrejón. https://web.archive.org/web/20250221142924/https://www.cervantesvirtual.com/obra/historia-de-las-indias--0/.

Laborde, Antonia. "La identidad puertorriqueña que renace tras el huracán." 14 Oct. 2018, https://web.archive.org/web/20250310151223/https://elpais.com/cultura/2018/10/14/actualidad/1539541922_257176.html.

Lalo, Eduardo. *Donde*. Tal Cual, 2005.

Lalo, Eduardo. *La inutilidad*. Corregidor, 2004.

Lalo, Eduardo. *La isla silente*. Isla Negra, 2002.

Laguerre, Enrique A. "Este ser inválido." *El Mundo*, 28 March 1974, p. 7A.

Laguerre, Enrique A. *La llamarada*. Editorial Cultural, 1994.

Laguerre, Enrique A. *Solar Montoya*. Editorial Cultural, 1967.

Lara, Ana-Maurine, et al. "Boricuir Trans-territorial Ecologies: Archipelagic Cimarronaje and Hemispheric Resurgence in Abya Yala." *CENTRO Journal*, vol. xxxv, no. 1 (spring 2023): 153–177.

Larkin, Brian. "The Politics and Poetics of Infrastructure." *Annual Review of Anthropology*, vol. 42 (2003): 327–343.

Lind-Ramos, Daniel. *El Viejo Griot—Una historia de todos nosotros*. MoMA PS1. Visited 20 May 2023.

Lloréns, Hilda. *Making Livable Worlds: Afro-Puerto Rican Women Building Environmental Justice*. U of Washington P, 2021.

Lloréns, Hilda et al. "Introduction: Centering Environmental, Ecological, Climate, and Energy Justice in Puerto Rico's Social Justice Praxis." *CENTRO Journal*, vol. xxxv, no. 1 (spring 2023): 7–21.

Lloréns Torres, Luis. "Canción de las Antillas." https://web.archive.org/web/20250310151848/https://ciudadseva.com/libros-de-luis-lopez-nieves/

Lloréns Torres, Luis. *Obras completas*. Instituto de Cultura Puertorriqueña, 1967.

López, Ivette. "Algo esconde paisajes: Julia de Burgos y el paisaje en la lírica del 30." *Revista de Estudios Hispánicos* (Puerto Rico), vol. 27, no. 2 (2000): 97–118.

López, Ramón. *Puerto Rico, USA: Historia de un país imaginario y otros ensayos*. Ediciones Huracán, 2000.

López-Baralt, Mercedes. "El huracán mítico de Palés Matos." *El Semanal*, 966, Sept. 2013. https://web.archive.org/web/20250310153019/https://www.jornada.com.mx/2013/09/08/sem-mercedes.html.

López-Baralt, Mercedes, editor. *Literatura puertorriqueña del siglo XX: Antología*. Universidad de Puerto Rico, 2004.

López Cantos, Ángel. *Los puertorriqueños: mentalidad y actitudes (Siglo XVIII)*. Ediciones Puerto, 2000.

López Marrero, T. y Castro Rivera, A. *Actividad ciclónica en Puerto Rico y sus alrededores: 1867 al 2017*. Centro Interdisciplinario de Estudios del Litoral, Universidad de Puerto Rico Mayagüez campus, 2018.

Lousley, Cheryl. *Ecocriticism*. 2020. https://web.archive.org/web/20250310152907/https://oxfordre.com/literature/display/https://doi.org/10.1093/acrefore/9780190201098.001.0001/acrefore-9780190201098-e-974

Luna, Noel. "Paisaje, cuerpo e historia: Luis Lloréns Torres." *La Torre: Revista de la Universidad de Puerto Rico*, vol. IV, no. 11: 53–78.

Maldonado Denis, Manuel. "Reflexiones en torno a un mito: La docilidad del puertorriqueño", Casa de Las Américas, XXI, (123), 1981: 137–146.

Manrique Cabrera, Francisco. *Historia de la literatura puertorriqueña*. Publicaciones Gaviota, 2010.

Marqués, René. "El puertorriqueño dócil. (Literatura y realidad psicológica)". *El puertorriqueño dócil y otros ensayos (1953-1971)*. 3rd ed. Editorial Antillana, 1977: 154–215.

Marqués, René. *La carreta*. Editorial Cultural, 1963.

Marrero Cabán, Sara R. "Denuncian que el gobierno está ajeno a las necesidades de la población sorda." https://web.archive.org/web/20220721205201/https:/www.primerahora.com/noticias/puerto-rico/notas/denuncian-que-el-gobierno-esta-ajeno-a-las-necesidades-de-la-poblacion-sorda/

Martínez-San Miguel, Yolanda. *Coloniality of Diasporas. Rethinking Intra-Colonial Migrations in a Pan-Caribbean Context*. Palgrave MacMillan, 2014.

Martorell, Antonio. *El velorio*. Ediciones R.I.P., 2010.

Matos Bernier, Félix. *Isla de arte*. Imprenta La Primavera, 1907.

Matos Paoli, Francisco. "A Don Pedro Albizu Campos." https://web.archive.org/web/20250221143917/https://www.poesi.as/fmp71010.htm.

Medina, Luis A. "Juracán: Guabancex's Favorite Weapon." http://emmy.uprrp.edu/lmedina/papers/juracan/index.html

Medrano Herrero, Pío. "Angustia, destrucción, pobreza y muerte. Los huracanes de 1615 y 1642 en Puerto Rico". *FOCUS*, vol. VII, no. I (2008): 19–32.

Meléndez, Edwin et al. "Post-Hurricane Maria Exodus from Puerto Rico and School Enrollment in Florida" CENTRO. Center for Puerto Rican Studies,

Hunter CUNY, 12-2017 https://web.archive.org/web/20250322130359/https://academicworks.cuny.edu/cgi/viewcontent.cgi?article=1006&context=cpr_pubs

Meléndez Muñoz, Miguel. *Cuentos del cedro*. Ediciones Rvmbos, 1967.

Meléndez-Badillo, Jorell. *Puerto Rico: Historia de una nación*. Planeta, 2024.

Méndez Ballester, Manuel. *Tiempo muerto. Tragedia en tres actos*. Ediciones Baldrich, 1940.

Mentz, Steve. "Shakespeare and the Blue Humanities." *SEL Studies in English Literature 1500-1900*, vol. 59, no. 2 (Spring 2019): 383–392. https://doi.org/10.1353/sel.2019.0018

Middeldyk, Rudolph Adams Van. *The History of Puerto Rico, from the Spaniard Discovery to the American Occupation*. Martin G. Brumbaugh, ed. Middletown, 2022, [originally published 1903] https://www.google.com/books/edition/The_History_of_Puerto_Rico/599tAAAAMAAJ?hl=en&gbpv=0

Millás, José Carlos. *Hurricanes of the Caribbean and Adjacent Regions 1492–1800*. Academy of the Arts and Sciences of the Americas, 1968.

Miner Solá, Edwin. *Historia de los huracanes en Puerto Rico*. First Book Publishing, 1995.

Miranda-Rodríguez, Edgardo. *La Borinqueña*. Somos Arte, LLC, 2016.

Miranda-Rodríguez, Edgardo, producer, *Ricanstruction. Reminiscing & Rebuilding Puerto Rico*. Somos Arte, LLC, 2018.

Mixer, Knowlton. *Porto Rico. History and Conditions*. The MacMillan Company, 1926.

MoMA PS1 / Programs / Daniel Lind-Ramos: *El Viejo Griot—Una historia de todos nosotros* Apr 20 – Sep 4, 2023. https://web.archive.org/web/20240914145419/https://www.momaps1.org/en/programs/178-daniel-lind-ramos

Montero, Mayra. *Aguaceros dispersos*. Tusquets Editores, 1996.

Moore, Jason W. "Who Is Responsible for the Climate Crisis?" 04 Nov. 2019. Blog. https://web.archive.org/web/20250322145212/https://www.maize.io/cultural-factory/what-is-capitalocene/

Morales, Iris. *Voices from Puerto Rico. Post-Hurricane Maria*. Red Sugarcane Press, 2019.

Morton, Timothy. *The Ecological Thought*. Harvard UP, 2010.

Murray-Román. Jeannine. "Audre Lorde's and Sofía Gallisá Muriente's Hurricane Diaries: Writing Strategies of Survival in Colonial Disaster." *Journal of West Indian Literature*, vol. 32, no. 1 (Nov. 2023): 19–39.

Museo de Arte Contemporáneo de Puerto Rico. *El Momento del Yagrumo*. https://artsandculture.google.com/story/owVRR57Q_mfGKA.

Museo de Arte de Puerto Rico. *The Memory of the Landscape*. https://artsandculture.google.com/story/the-memory-of-the-landscape-museo-de-arte-de-puerto-rico/sgUB1Rn5YY2L1Q?hl=en.

Naranjo Orovio, Consuelo, director. *Historia de las Antillas. Vol. IV. Puerto Rico.* Consejo Superior de Investigaciones Científicas, Ediciones Doce Calles, 2012.

National Academy of Sciences and the Royal Society. *Climate Change: Evidence and Causes: Update 2020.* National Academies P, 2020, https://doi.org/10.17226/25733.

National Oceanic and Atmospheric Administration (NOAA). Online databases as noted. coast.noaa.gov/hurricanes.

National Oceanic and Atmospheric Administration (NOAA). "Project STORMFURY" https://www.aoml.noaa.gov/hrd/hrd_sub/sfury.html

National Oceanic and Atmospheric Administration (NOAA). "Saffir-Simpson Hurricane Wind Scale." www.nhc.noaa.gov/aboutsshws.php.

Navarro Aquino, Xavier. *Velorio.* Harper Collins Español, 2022.

Neely, Wayne. *The Greatest and Deadliest Hurricanes of the Caribbean and the Americas.* iUniverse, 2019.

Negrón-Muntaner, Frances and Magali García Ramis. "Interview." *Hispamérica*, 22, No. 64/65 (Apr.-Aug., 1993): 89–104. Stable URL: https://www.jstor.org/stable/20539689

Nixon, Rob. *Slow Violence and the Environmentalism of the Poor.* Harvard UP, 2011.

Ojeda Matos, Glorynel. "Las maravillas del Caño Martín Peña." *Verdor que clama*, Carmen M. Rivera Villegas et al, editors. Ediciones del Flanboyán, 2022, pp. 55–62.

Oliva, Frank, et al. *Paleotempestology Database for the Western North Atlantic Basin. The Holocene.* 2018. https://doi.org/10.1177/0959683618782598.

Oliver-Smith, Anthony, and Susanna M. Hoffman. 'Why Anthropologists Should Study Disaster' *Catastrophe and Culture: The Anthropology of Disaster*, Susanna Hoffman and Anthony Oliver-Smith, editors. School of American Research P, 2002: 3–22.

Oppermann, Serpil. "Storied Seas and Living Metaphors in the Blue Humanities." *Configurations*, vol. 27, no. 4 (fall 2019): 443–461. https://doi.org/10.1353/con.2019.0030

Orsanic, Lucía and Jorge Fusaro Martínez, editors. *Pa la posteridá. Antología sobre el paso del huracán María por Puerto Rico.* Ediciones del Flanboyán, 2018.

Ortíz, Fernando. *El huracán. Su mitología y sus símbolos.* Fondo de Cultura Económica, 1947.

Pacific Coastal and Marine Science Center. "Hokusai under the Wave off Kanagawa." www.usgs.gov/media/images/hokusai-under-wave-kanagawa.

Palés Matos, Luis. *Tuntún de pasa y grifería.* Editorial de la Universidad de Puerto Rico, 2003.

Palés Matos, Luis. *Selected Poems: Poesía Selecta.* Arte Público Press, 2000.

Paravisini-Gebert, Lizabeth. "Endangered Species. Caribbean Ecology, and the Discourse of the Nation." *Displacements and Transformations in Caribbean Cultures*, Lizabeth Paravisini-Gebert and Ivette Romero-Cesareo, editors. Florida UP, 2008: 8–23.

Pastrana, Francisco, Mateo Cavailhon y Alejandrina Benítez, et al. *Aguinaldo puertorriqueño*. 1843. *bdh-rd.bne.es/viewer.vm?id=0000125125&page=1.*

Pedreira, Antonio S. *Insularismo*. Editorial Edil, 1971.

Pérez Losada, José. *El manglar: novela puertorriqueña* (*2a. parte de La Patulea*). Puerto Rico, Boletín Mercantil, 1909. https://www.google.com/books/edition/El_manglar/EjhEAQAAMAAJ?hl=en&gbpv=0

Pérez Martínez, Bettina. "Puerto Rican Visual Arts in the Era of Superstorms: An Ecocritical and Decolonial Analysis of the Work of Frances Gallardo and Lionel Cruet." *CENTRO Journal*, vol. xxxv, no. 1 (spring 2023): 123–151.

Picó, Fernando. *Puerto Rico Inside and Out: Changes and Continuities*. Markus Weiner Publishers, 2008.

Picó, Fernando. *Historia general de Puerto Rico*. 5th ed. Huracán, 1990.

Picó, Fernando. *History of Puerto Rico. A Panorama of its People*. Updated ed. Markus Weiner Publishers, 2017.

Picó, Rafael. *Geografía de Puerto Rico. Parte I geografía física*. Editorial Universitaria, 1954.

Ponce de León, Juan. "Letter to the Governor of the Indies, Nicolás de Ovando." 1509. Luis A. Salivia, *Historia de los temporales de Puerto Rico:* p. 4.

Ponce de León II, Juan (Troche) and Antonio de Santa Clara. *Memoria de Melgarejo*. 1582. http://smjegupr.net/wp-content/uploads/2012/05/Memoria-de-Johan-de-Melgarejo.pdf

Pörtner, Hans-O. et al. "2020 Intergovernmental Panel on Climate Change (IPCC) Sixth Assessment Report. Impacts, Adaptation and Vulnerability." *United Nations' Intergovernmental Panel on Climate Change*. www.ipcc.ch/report/ar6/wg2/

Poupeye, Veerle. *Caribbean Art*. Thames and Hudson, 1998.

Povinelli, Elizabeth A. "The Rhetorics of Recognition in Geontopower." *Philosophy & Rhetoric*, vol. 48, no. 4, Special Iss: "The Rhetorical Contours of Recognition" (2015): 428–442.

Price Lee, Mary, and Richard S. Lee. *Natural Disasters and Post-Traumatic Stress Disorder*. The Rosen Publishing Group, 1996.

Puerto Rico Hurricanes Map. www.usgs.gov/media/images/puerto-rico-hurricanes-map.

"Puerto Rico v. Sanchez Valle." Oyez. https://web.archive.org/web/20241121080028/https://www.oyez.org/cases/2015/15-108

Quintana, Hilda E., et al. *Personalidad y literatura puertorriqueñas*. Editorial Plaza Mayor, 1996.

Ramos Collado, Lilliana. "'¡Allá viene el temporal': Representación de la catástrofe entre lo culto y lo popular." 1 Feb. 2012. https://web.archive.org/web/20250322162855/https://bodegonconteclado.wordpress.com/2012/01/23/alla-viene-el-temporal-representacion-de-la-a-catastrofe-entre-lo-culto-y-lo-popular/ /.

Ramos Collado, Lilliana. "En busca del paraíso... perdido." 24 Jan. 2012. https://web.archive.org/web/20250322163340/https://bodegonconteclado.wordpress.com/tag/paraiso/

Ramos Isabel and Kisha Tikina Burgos. *Antígona frente al mar* (version1.8), 2023. Manuscript cited by permission of the authors.

Rappaport, Edward N, and Jose Fernandez-Partagas. "The Deadliest Atlantic Tropical Cyclones, 1492–1996". 28 May 1995, NOAA Technical Memorandum NWS NHC 47. Updated 22 April 1997 by Jack Beven. www.nhc.noaa.gov/pastdeadly.shtml

Renan, Ernest. "What is a Nation?" 1882. Ethan Rundell, translator. Presses-Pocket, 1992. https://ucparis.fr/files/9313/6549/9943/What_is_a_Nation.pdf.

Rivera, Danielle Zoe. "Disaster Colonialism: A Commentary on Disasters beyond Singular Events to Structural Violence." *International Journal of Urban and Regional Research*. Vol. 46, Iss. 1 (Jan. 2022): 1–156. https://doi.org/10.1111/1468-2427.12950

Rivera, Juan Pablo. *La hermosa carne. El cuerpo en la poesía puertorriqueña actual*. Iberoamericana Vervuert, 2021.

Rivera, Nelson. *Inconformes y disidentes. Arte puertorriqueño en el siglo XXI*. Editorial Isla Negra, 2023.

Rivera Chevremont. Evaristo. *Canto de mi tierra*. Editorial Universitaria, 1971.

Rivera de Álvarez, Josefina. *Literatura puertorriqueña. Su proceso en el tiempo*. Ediciones Partenón, 1983.

Rivera de Álvarez, Josefina and Manuel Álvarez Nazario. *Antología general de la literatura puertorriqueña. Volume I*. Ediciones Partenón, 1982.

Rodríguez Cotto, Sandra. *Bitácora de una transmisión radial*. Trabalis, 2018.

Rodríguez Cruz, Rafael. *El huracán y la subjetividad antillana*. Ediciones Gran Sempié, 2022.

Rodríguez de Tió, Lola. "A Puerto Rico," "La borinqueña" https://web.archive.org/web/20250322163724/https://ciudadseva.com/texto/la-borinquena/

Rodríguez de Tió, Lola. "Borinquen, nido de flores." https://web.archive.org/web/20250322164017/https://ciudadseva.com/texto/borinquen-nido-de-flores/

Rodríguez de Tió, Lola. *Claros y nieblas*. Tipografía Comercial, 1885. https://www.google.com/books/edition/Claros_y_nieblas/gM0pAAAAYAAJ?hl=en&gbpv=0.

Rodríguez Juliá, Edgardo. *Campeche o los diablejos de la melancolía*. Instituto de Cultura Puertorriqueña, 1986.

Rodríguez-Juliá, Edgardo. *La noche oscura del niño Avilés*. Biblioteca Ayacucho, 2002.

Rodríguez-Juliá, Edgardo. *San Juan, ciudad soñada*. Editorial Tal Cual, 2005.

Rodríguez-Juliá, Edgardo. *The Renunciation: A Novel*. Andrew Hurley, translator. Four Walls Eight Windows, 1997.

Rodríguez Martinó, Graciela. "'¿Quéslaque? Es que la…' Exposición de Antonio Martorell en memoria a los muertos de María." *80grados*, 17 Aug. 2018. https://web.archive.org/web/20250322164833/https://www.80grados.net/web/20250322164343/https://www.80grados.net/queslaque-es-que-la-exposicion-de-antonio-martorell-en-memoria-a-los-muertos-de-maria/?__cf_chl_rt_tk=rzuKWhpAEw1pqL8FWceOZF9jEBERzwbZEuWuZBZD0rU-1742662113-1.0.1.1-gmmtNWmidtM_KdJ5L55O761SqXSc0JfhmygGXR6CCwI.

Rogers, Charlotte. "The Environmentalism of Poor Women of Color in Mayra Santos-Febres's *Nuestra Señora de la Noche*." Kressner, Ilka, Ana María Mutis and Elizabeth M Pettinaroli, editors. *Eco-realities and Slow Violence in Latin America and the Latinx World*. Routledge (2020): 180–196.

Rosario Natal, Carmelo. *El puertorriqueño dócil: Historia, pasión y muerte de un mito*. U of Virginia P, 1987.

Rozario, Kevin. *The Culture of Calamity. Disaster and the Making of Modern America*. Chicago UP, 2007.

Rueckert, William. "Literature and Ecology: An Experiment in Ecocriticism." Glotfelty, Cheryll and Fromm, Harold, editors. *The Ecocriticism Reader: Landmarks in Literary Ecology*. University of Georgia Press (1996): 105–23.

Salas Rivera, Raquel. *While they sleep[1 1] under the bed is another country*. Birds LLC, 2019.

Salivia, Luis A. *Historia de los temporales de Puerto Rico*. Imprenta La Milagrosa, 1950.

Sanabria Santalíz. "Después del huracán." Efraín Barradas, editor. *Apalabramiento. Diez cuentistas de hoy*. Ediciones del Norte, 1983: pp. 156–169.

Sánchez, Luis Rafael. *La guaracha del Macho Camacho*. De la Flor, 1976.

Sánchez, Luis Rafael. *El corazón frente al mar*. Publicaciones Gaviota, 2021.

Santiago, Esmeralda. *Las madres*. Alfred A. Knopf, 2023.

Santiago Álvarez Carolina E. *Mitología taina*. Mitología _de_ los_ Tainos de Puerto Rico.pdf. N.d. 802-13-7925, HIST 3241 Sección 050.

Santos Febres, Mayra. *Antes que llegue la luz*. Planeta, 2021.

Santos Febres, Mayra. *Huracanada*. Trabalis Editores, 2018.

Santos-Febres, Mayra and Daniel Darío Carrero. *En el ojo del huracán*. La Otra Orilla, 2011.

Schecter Patricia A., "Feminist Historiography, Anti-imperialism, and the Decolonial." *Empire's Twin. U.S. Anti-imperialism from the Founding Era to the Age of Terrorism.* Ian Tyrrell and Jay Sexton, editors. Cornell UP, 2015. pp. 153–166.

Schwartz, Steward B. *Sea of Storms. A History of Hurricanes in the Greater Caribbean from Columbus to Katrina.* Princeton UP, 2015.

Sheets, Bob and Jack Williams, *Hurricane Watch: Forecasting the Deadliest Storms on Earth.* Vintage, 2001.

Schnabel, Rose. "Estudio de la UPR sugiere que los primeros pobladores de Puerto Rico llegaron 3,000 años antes de lo establecido." *El Nuevo Día*, 15 July 2023.

Sigüenza y Góngora, Carlos de. *Infortunios de Alonso Ramírez.* Madrid. 1690. 1902. https://archive.org/details/infortuniosdealo00siguoft/page/n2/mode/1up?ref=ol&view=theater

Smith Silva, Dorsía. *Inheritance of Drowning.* CavanKerry, 2024.

Sultana, Farhana. "Suffering for Water, Suffering from Water: Emotional Geographies of Resource Access, Control, and Conflict." *Geoforum* vol. 42, no. 2 (March 2011): 163–172.

Tapia y Rivera, Alejandro, et al. *Biblioteca histórica de Puerto-Rico: que contiene varios documentos de los siglos XV, XVI, XVII y XVIII.* Marquez: 1854. https://www.loc.gov/item/07041355/

Tapia y Rivera, Alejandro, et al. *El pirata Cofresí.* Puerto eBooks, 2015.

Tapia y Rivera, Alejandro, et al. *Mis memorias.* https://collections.ctdigitalarchive.org/node/48325

Tapia, Alejandro., Rivera (sic). Complete Works of Alejandro Tapia Y Rivera. United States: Gordon Press Publishers, 1979. https://www.google.com/.

Tirado de Delucca, Elba M. *Historia de Puerto Rico, siglo XX. Desde la presencia de Estados Unidos de América en Puerto Rico en 1898 hasta el año 1996.* [S.l.: s.n.], 1997.

Toro Soler, Ricardo. *Huracán. Leyenda puertorriqueña.* Tipografía la Libertad, 1897.

Trexler, Adam and Adeline Johns-Putra. "Climate Change in Literature and Literary Criticism." *WIREs Climate Change* vol. 2, iss. 2 (24 February 2011): 185–200. https://doi.org/10.1002/wcc.105.

Tufiño, Rafael and Lorenzo Homar. *12 grabados: La Plena.* Editorial Caribe, 1955.

Ubeda y Delgado, Manuel. *Isla de Puerto Rico. Estudio histórico, geográfico y estadístico de la misma.* Boletín, 1878. https://web.archive.org/web/20250322165456/https://issuu.com/adelantereunificacionistas/docs/isla_de_puerto_rico.

University of Rhode Island. School of Oceanography. "Ecosystem Perspective: What Can a Hurricane Do to the Environment?" www.hurricanescience.org/.

Valcárcel, Xavier. *Aterrizar no es regreso*. Ediciones Alayubia, 2019.

Valladares de Sotomayor, Antonio. *Historia Geográfica, Civil y Política de la Isla de San Juan Bautista de Puerto Rico (1788)* España, 1788. https://web.archive.org/web/20250322165702/https://issuu.com/adelantereunificacionistas/docs/historiageogrf00abba.

Varela, Julio Ricardo. "This is an Exhausting Time to Be Puerto Rican." *The Washington Post*, 04 Feb. 2020. https://www.washingtonpost.com/outlook/2020/02/04/this-is-an-exhausting-time-be-puerto-rican/.

Vega, Ana Lydia. *Encancaranublado y otros cuentos de naufragio*. 4th ed. Editorial Antillana, 1990.

Vega, Ana Lydia. *El tramo ancla*, editor. Editorial de la Universidad de Puerto Rico, 1989.

Vega, Ana Lydia. *Falsas crónicas del Sur*. Editorial de la Universidad de Puerto Rico, 1991.

Vega, Ana Lydia. "Marejada de los muertos." *El Nuevo Día*. 3 July 2011.

Vega, Ana Lydia. *Mirada de doble filo*. Editorial de la Universidad de Puerto Rico, 2008.

Vermuelen, Pieter. *Literature and the Anthropocene*. Routledge, 2020.

Veronesi, Mariangela, et al. "Community-led Development and Collective Land Tenure for Environmental Justice: The Case of the Caño Martín Peña Community Land Trust, Puerto Rico." *International Journal of Urban Sustainable Development*, 2022, https://doi.org: https://doi.org/10.1080/19463138.2022.2096616

Vidal, Teodoro. *Oraciones, conjuros y ensalmos en la cultura popular puertorriqueña*. Ediciones Alba, 2010.

Vignoli, Alessia. "Les catastrophes naturelles dans les romans caribéens d'expression française." Il Tolomeo, Vol. 22 (Dec. 2020): 149–167.

Vivas, José Luis. *Historia de Puerto Rico*. Las Americas Publishing Co, 1962. https://openlibrary.org/books/OL24634683M/Historia_de_Puerto_Rico.

Wagenheim, Kal. *Porto Rico. A Profile*. Second ed. Praeger Publishers. 1970.

Wallace, E.J. et al. "Intense Hurricane Activity over the Past 1500 Years at South Andros Island, The Bahamas." *Paleoceanography and Paleoclimatology*, vol. 34, issue 11 (2019): 1761–1783. https://web.archive.org/web/20250322170545/, https://agupubs.onlinelibrary.wiley.com/, https://doi.org/10.1029/2019PA003665.

Waymer, Jim. "Here's Why Scientists Say No. Don't Ever Underestimate the Power of a Storm." *Florida Today*, 5 Sept. 2019 https://www.floridatoday.com/

Wehner, Michael F. and James P. Kossin. "The Growing Inadequacy of an Open-Ended Saffir–Simpson Hurricane Wind Scale in a Warming World." *PNAS*,

121 (7) e2308901121, Feb. 5, 2024. (https://orcid.org/0000-0001-5991-0082 mfwehner@lbl.gov Info & Affiliations) https://doi.org/10.1073/pnas.2308901121

Whang, Oliver. "We've Run out of Hurricane Names. What Happens Now?" 22 Sept. 2020. https://www.nationalgeographic.co.uk/environment-and-conservation/2020/09/weve-run-out-of-hurricane-names-what-happens-now.

Wood, Gillen D'Arcy. "Introduction. Eco-Historicism." *The Journal for Early Modern Cultural Studies*. vol. 8, no. 2 (fall/winter 2008): 1–7.

Woodruff, Jonathan D., et al. "Reconstructing Relative Flooding Intensities Responsible for Hurricane-Induced Deposits from Laguna Playa Grande, Vieques, Puerto Rico." *Geology*, vol. 36, no. 5 (2008): 391–394. https://doi.org/10.1130/G24731A.1

Wordsworth, William. *The Prelude: The Four Texts (1798, 1799, 1805, 1850)*. Penguin Books Limited, 2004.

Yaeger, Patricia. "Editor's Column: Sea Trash, Dark Pools, and the Tragedy of the Commons." *PMLA*, vol. 125, no. 3 (2010): 523–45. www.jstor.org/stable/25704454. Accessed 25 May 2016.

Yaeger, Patricia. "Introduction: Dreaming of Infrastructure." *PMLA*, vol. 122, no. 1 (2007): 9–26. https://doi.org/10.1632/pmla.2007.122.1.9.

Yuan, Yuan. "Contextualizing Climate Change Fiction and Anthropocene Criticism in the 21st Century: An Interview with Adeline Johns-Putra." *Foreign Literature Studies/Wai Guo Wen Xue Yan Jiu*, vol. 44, no. 3 (June 2022): 1–17.

Zambrana, Rocío. *Colonial Debts*. Duke UP, 2021.

Zeno Gandía, Manuel. *La Charca*. https://web.archive.org/web/20250322170850/http://maestrojgcv.weebly.com/uploads/7/5/7/0/7570259/la_charca.pdf

Zhong, Raymond. "Geologists Make It Official: We're Not in an 'Anthropocene' Epoch." *The New York Times*, 20 March 2024. https://www.nytimes.com/2024/03/20/climate/anthropocene-vote-upheld.html?action=click&pgtype=Article&state=default&module=styln-climate&variant=show&region=MAIN_CONTENT_1&block=storyline_levelup_swipe_recirc.

INDEX

M. Acosta Cruz, *An Ecocritical Study of Puerto Rican Culture*,
https://doi.org/10.1007/978-3-032-04044-2

FSC
www.fsc.org
MIX
Papier | Fördert
gute Waldnutzung
FSC® C083411

Kolibri 360 GmbH
Ferdinand-Jühlke-Straße 7
99095 Erfurt, Deutschland
produktsicherheit@kolibri360.de